Algebra: A Complete Course

Second Edition

Module A – Instructor's Guide with Detailed Solutions for Progress Tests

Handwritten Solutions by Donna Freiburger

VideoText *Interactive*

Algebra: A Complete Course
Module A – Instructor's Guide with Detailed Solutions for Progress Tests, Second Edition

Send all inquiries to:
VideoText*Interactive*
P.O. Box 19761
Indianapolis, IN 46219

ISBN 1-887891-70-6
8 9 10 11 12 13 14 15 - RPInc - 12 11 10 09 08 07

Letter from the author . . .

Congratulations on your decision to include the VideoText*Interactive* Algebra course in your educational resources. You have chosen a delivery system that is allowing students nationwide to enjoy remarkable success in a variety of settings. I'm sure you will continue to be satisfied with your choice as you see your students develop a strong understanding of Algebra concepts, and an increased confidence in their own mathematical ability.

Please be sure to carefully review the program overview on pages iv and v, as well as the scope and sequence on pages vi and vii. I wrote it to help you, as the instructor, to understand the dynamic quality of the instructional process and to afford you the most efficient use of the materials. In fact, if your students do not involve themselves personally and actively in the learning process – pausing the tapes frequently, answering questions before the video lessons do, and engaging in analysis – their understanding of the concepts will not be as deep as it should be and those concepts will not be remembered as long.

Again, we at VideoText*Interactive* are very happy you have taken this step and we look forward to hearing your comments and suggestions. We especially want to know about the successes you have enjoyed as a result of using the program. Good luck!

Sincerely,

Thomas E. Clark
Author

Table of Contents

Instructional Aids

Course Schematic iii

Program Overview iv

Scope and Sequence Rationale vi

Detailed Solutions for Progress Tests

Unit I – The Structure of Mathematics

Part A – Mathematics as a Language

LESSON 1 Mathematical Parts of Speech
- Quiz A 1
- Quiz B 3

LESSON 2 Mathematical Expressions
- Quiz A 5
- Quiz B 7

LESSON 3 Translation of Mathematical Symbols
- Quiz A 9
- Quiz B 11

Part B – Further Investigation of Number Symbols

LESSON 1 The Development of Our Number System
- Quiz A 13
- Quiz B 15

LESSON 2 Fraction Forms and Decimal Forms] combined
LESSON 3 Changing Fraction Forms to Decimal Forms] combined
- Quiz A 17
- Quiz B 19

LESSON 4 Changing Decimal Forms to Fraction Forms] combined
LESSON 5 Percent] combined
- Quiz A 21
- Quiz B 23

LESSON 6 Primes, Composites, and Factoring] combined
LESSON 7 Least Common Multiple] combined
LESSON 8 Greatest Common Factor] combined
- Quiz A 25
- Quiz B 27

Part C – Further Investigation of Operation Symbols

LESSON 1 Order of Operations
LESSON 2 Properties of Operations
LESSON 3 Properties of Operations with Special Numbers] combined
Quiz A 29
Quiz B 31
LESSON 4 Operations with Fractions – Multiplication
LESSON 5 Operations with Fractions – Addition and Subtraction
LESSON 6 Operations with Fractions – Division] combined
Quiz A 33
Quiz B 35
LESSON 7 Operations with Decimals
Quiz A 37
Quiz B 39
LESSON 8 Operations with Signed Numbers – Vectors and Absolute Value
LESSON 9 Operations with Signed Numbers – Addition
LESSON 10 Operations with Signed Numbers – Subtraction] combined
Quiz A 41
Quiz B 43
LESSON 11 Operations with Signed Numbers – Multiplication and Division
Quiz A 45
Quiz B 47

Part D – Further Investigation of Relation Symbols

LESSON 1 Order of Numbers and the Number Line
Quiz A 49
Quiz B 51
LESSON 2 Properties of Equality
LESSON 3 Properties of Inequality] combined
Quiz A 53
Quiz B 55

Part E – Mathematical Models

LESSON 1 The Mathematics of Sets
Quiz A 57
Quiz B 59
LESSON 2 The Mathematics of Functions
Quiz A 61
Quiz B 63

Unit Test Form A 65

Unit Test Form B 71

Index 77

ALGEBRA - A COMPLETE COURSE

STRUCTURE OF MATHEMATICS (Unit I)

TAPE 1

- A. Mathematics as a Language
  - Parts of Speech
  - Expressions
  - Translations
- B. More on Numbers
  - Number System
  - Fraction & Decimal Forms)
  - Equivalent Forms - 1
  - Equivalent Forms - 2
  - Percent
  - Primes and Composites
  - Least Common Multiple
  - Greatest Common Factor

TAPE 2

- C. More on Operations
  - Order
  - Properties - 1
  - Properties - 2
  - Fractions - 1
  - Fractions - 2
  - Fractions - 3
  - Decimals
  - Signed Numbers - 1
  - Signed Numbers - 2
  - Signed Numbers - 3
  - Signed Numbers - 4

TAPE 3

- D. More on Relations
  - Order
  - Properties - 1
  - Properties - 2
- E. Models
  - Sets
  - Functions

FIRST DEGREE RELATIONS

One Placeholder (Unit II)

TAPE 4

- A. Basic Relations
  - Solutions
  - Making Zeros
  - Making Ones
  - Combinations
- B. Complications
  - Grouping Symbols
  - Like Terms on the Same Side
  - Placeholders on Both sides
  - Combinations
- C. Special Cases
  - No Solution
  - Infinite Solutions
- D. Systems
  - "And"
  - "Or"
  - Absolute Value - 1
  - Absolute Value - 2
  - Absolute Value - 3

TAPE 5

- E. Problem Solving
  - Setup
  - "Numbers"
  - "Consecutive Integers"
  - "Age"
  - "Geometric Figures"
  - "Motion"
  - "Percent"

Two Placeholders (Unit III)

TAPE 6

- A. Solution Sets
  - Equations
  - Inequalities
  - Graphing Terms
  - $y = mx$
  - $y = mx + b$
  - Intercepts
- B. Special Cases
  - $y =, >, < a$
  - $x =, >, < a$
  - Absolute Value

TAPE 7

- C. Relations from Solutions
  - Given Slope and y-intercept
  - Given Slope and One Solution
  - Given Two Solutions
  - Special Cases
- D. Systems
  - Graphic - 1
  - Graphic - 2
  - Algebraic - 1
  - Algebraic - 2

TAPE 8

- E. Special Cases
  - No solution
  - Infinite Solutions
- F. Problem Solving
  - Setup
  - "Numbers"
  - "Age"
  - "Geometric Figures"
  - "Motion"
  - "Percent"
  - "Mixture"

Three Placeholders (Unit IV)

TAPE 9

- A. Solution Sets
  - One Sentence
  - Two Sentences
  - Three or More Sentences
- B. Special Cases
  - No Solution
  - Infinite Solutions
- C. Problem Solving
  - "Numbers"
  - "Age"
  - "Geometric Figures"
  - "Mixture"

SECOND DEGREE RELATIONS (or higher)

Polynomials (Unit V)

TAPE 10

- A. Exponent Notation
  - Terminology
  - Operations - 1
  - Operations - 2
  - Special Cases
  - Scientific Notation
- B. Polynomials
  - Terminology - 1
  - Terminology - 2
  - Operations - 1
  - Operations - 2
  - Operations - 3

TAPE 11

- C. Solving Relations
  - Principle of Zero Products
  - Common Factor
  - Difference of Squares
  - Perfect Square Trinomial
  - General Trinomial
  - Four or More Terms
  - Sum or Difference of Cubes
  - General Strategy
  - Synthetic Division
  - Literal Equations

TAPE 12

- D. Problem Solving
  - "Numbers"
  - "Consecutive Integers"
  - "Geometric Figures"
  - "Formula"

Algebraic Fractions (Unit VI)

TAPE 13

- A. Operations
  - Simplifying
  - Multiplying
  - Dividing
  - Adding or Subtracting
  - Complex
- B. Solving Relations
  - Equations - 1
  - Equations - 2
  - Inequalities
  - Literal Equations
- C. Problem Solving
  - "Fractions"
  - "Work"
  - "Motion"
  - "Direct Variation"
  - "Inverse Variation"
  - "Mixed Variation"

RATIONAL DEGREE RELATIONS (Unit VII)

TAPE 14

- A. Rational Exponents
  - Radicals
  - "x^{th}" roots
- B. Operations
  - Multiplying
  - Simplifying
  - Dividing
  - Adding or Subtracting
  - In Polynomials
  - Rationalizing
- C. Solving Relations
  - One Radical
  - Two Radicals
- D. Problem Solving
  - "Distance"
  - "Formula"
- E. Complex Numbers
  - Definition
  - Adding or Subtracting
  - Multiplying
  - Dividing

SUMMARY OF SECOND DEGREE RELATIONS*

Quadratic (Unit VIII)

TAPE 15

- A. Solving from Standard Form
  - $a, b, \text{or } c = 0$
  - $a, b, c \neq 0$
  - Quadratic Formula
  - Checking Solutions
  - Inequalities (0:05:33)
- B. Relations Quadratic In Form
  - Higher Order to Lower Order > 0
  - Lower Order to Higher Order > 0
  - Integer Order < 0
- C. Problem Solving
  - "Geometric Figures"
  - "Pythagorean Theorem"
  - "Work"
  - "Motion"

Conic Sections (Unit IX)

TAPE 16

- A. Parabolas
  - Origins
  - Standard Form - 1
  - Standard Form - 2
  - Standard Form - 3
  - Standard Form - 4
  - Intercepts
- B. Circles
  - Standard Form
  - Non-standard Form
- C. Ellipses
  - Standard Form
  - Non-standard Form
- D. Hyperbolas
  - Standard Form
  - Non-standard Form
- E. Systems
  - One 1st Degree and One 2nd Degree
  - Two 2nd Degree
- F. Problem Solving
  - "Numbers"
  - "Geometric Figures"

LITERAL DEGREE RELATIONS (Unit X)

TAPE 17

- A. Exponential Functions
  - Functions of x
  - Functions of y
- B. Logarithmic Functions
  - Translating
  - Graphs of Solution Sets
- C. Operations
  - Properties
  - Finding Logarithms)
  - Computation
- D. Solving Equations
  - Exponential
  - Logarithmic

* This would logically follow ALGEBRAIC FRACTIONS (Unit VI) except that a knowledge of roots and radicals is necessary for a comprehensive summary.

Program Overview

The VideoText Algebra program addresses two of the most important aspects of mathematics instruction. First, the **inquiry-based video format contributes to the engaging of students** more personally in the concept development process. Through the frequent use of the pause button, you, as the instructor, can virtually require interaction and dialogue on the part of your student. As well, students who work on their own, can "simulate" having an instructor present by pausing the tape every time a question is asked and trying to answer it correctly before continuing. Second, **each incremental concept is explored in detail**, using no shortcuts, tricks, rules, or formulas, and no step in the process in ignored. As such, the logic and the continuity of the development assures students that they understand completely. Subsequently, learning is more efficient, and all of the required concepts (topics) of the subject can be covered with mastery. Of course, the benefits of these efforts can be seen even more clearly in a description of a typical session as follows:

After a brief 2 or 3 sentence introduction of the concept to be considered, usually by examining the description and the objective given at the beginning of the video lesson, you and your student can begin. **You should pause the tape frequently**, usually every 15-20 seconds (or more often if appropriate), to engage your student in discussion. This means that for a 5-10 minute VideoText lesson, it may take 15-20 minutes to finish developing the concept. **Dialogue is a cornerstone**. In addition, during this time, your student should probably not be allowed to take notes. He or she must not have their attention divided, or they risk missing important links. Neither should you be dividing your attention by looking at notes or writing on a note pad or an overhead projector. Everyone is concentrating on concept development and understanding. Please understand that a student who is accustomed to working alone, or can be motivated to study independently, has, with the VideoText, a powerful resource to explore and master mathematical concepts by simulating the dialogue normally encountered with a "live" instructor. And, because of the extensive detail of the explanations along with the computer generated graphics and animation, students are never shortchanged when it comes to the insight necessary to fully comprehend.

Once the concept is developed and the VideoText lesson is completed, you can then employ the course notes to review, reinforce, or to check on your student's comprehension. These course notes are exact replications of the content that was viewed in the VideoText lesson, illustrating the same terms, problems, numbers, and logical sequences. In fact, at this time, if your student needs a little more help, he or she can use these course notes while viewing the lesson again, using them as a guide through the reexamination of the concept. **The key here is that students concentrate on understanding first, and take care of documentation later.**

Please understand that it is not the intent of the program to let the VideoText lesson take the place of instruction. **Actually, the video should never tell your student anything that hasn't been considered or discussed (while the tape is paused), and it should never answer questions that have not already been resolved**. As such, it becomes a "new breed" of overhead projector whereby you, as the teacher, or your student working alone, can "write" on this overhead simply by pressing the "play" button. This is a critical point to be understood and should serve to help you examine all of the materials and strategies from the proper perspective.

Finally, your student can begin to do some work independently, either by your introduction of additional examples from the WorkText, or by the student immediately going to the WorkText on his or her own. The primary feature of the WorkText, beside providing problem banks with which students can work on mastery, is that **objectives are restated, important terms are reviewed**, and **additional example are considered**, in noticeable detail, taking students, once again, through the logic of the concept development process. The premise here is simple. When students work with an instructor, whether doing exercises on their own or working through them with other students, they are concentrating more on "how to do" the problems. Then, when they leave the instructor, they simply don't take the discussion of the concept with them. The goal of the program is to provide a resource which will **help students "relive" the concept development** on their own, whether for review or for additional help. That is the focus of the Student WorkText.

In addition, there are detailed Solutions Manuals, which afford the opportunity to **check work** and **engage in error analysis**. Of course, there are also quizzes, unit tests, cumulative reviews, and final examinations to help you to further assess your student's progress. In fact, the assessment package often utilizes **open-response questions** which require the learners to state, in writing, their understanding of the concept. This often reveals much more about a student's understanding of a concept than just checking to see if an answer on a test is correct.

As you can see, the highly interactive quality of this program, at a personal level, affords students a much greater opportunity than usual to grow mathematically and develop confidence in their ability. As well, they can review the video lessons as often as they wish to further ground that understanding.

Scope and Sequence Rationale

There are two basic premises which drive concept development in Algebra, and these two essentials shape the logical scope and sequence of algebraic content.

First, it is generally understood that the **study of Algebra is the study of relations**. In the same way that Geometry focuses on spatial concepts, and Calculus is concerned with rates of change, Algebra is a comprehensive exploration of mathematical relationships, including both equations and inequalities. As such, no treatment of Algebra should ever separate equations from inequalities, especially when it utilizes a format which addresses them in different "chapters." In fact, a true adherence to the National Council of Teachers of Mathematics (NCTM) standards requires us to deal specifically with functions, and we know that the set of functions is a subset of the set of relations, without regard initially to the differences between equations and inequalities. Therefore, in this course, **equations and inequalities are studied together** and distinctions are made only when necessary to clarify functional differences. As an aside, documentation exists to show that students generally have little or no trouble working with all types of relations at the same time and, in fact, understand the logic of studying them together.

The second premise is that **the concepts of Algebra develop by degrees**. This means, of course, that relations of first degree should be mastered first. In fact, as instructors, we all understand that relations of any degree other than one must be "reduced" to relations of first degree or "factored" into linear or first degree factors before they can be resolved. The impact of this understanding on the scope and sequence of Algebra content is to organize the various types of relations by degree. In this course, first degree relations are examined exhaustively before higher order relations are encountered. Unit II deals with first-degree relations with one variable. Unit III then addresses first-degree relations with two variables. Unit IV considers first-degree relations with three or more variables. The idea here is to help students **master first degree relations before moving on to relations of other degrees** (or orders). This is not only more mathematically correct than the traditional treatment, but it allows students to reinforce more efficiently, one-variable concepts by immediately moving to two-variable concepts, and then to concepts involving three or more variables. And we all know that a system of relations with three variables is resolved using the same approach as a system with only two variables.

Moving on to Unit V, students quickly review exponent notation, including the various properties of powers and operations with powers and investigate relations with integral degrees of 2 or higher. Unit VI continues this exploration with a focus on algebraic fractions, in which negative, integral exponents make a prominent appearance. In Unit VII, fractional exponents are introduced,which obviously pave the way for a study of radicals and roots. This, of course, is the seed from which rational-degree relations develop or, as they are more commonly called, relations with radicals in them.

Then, after a review of second-degree relations with one variable (Unit VIII – The Quadratic Relations) and two variables (Unit IX – The Conic Sections), the study of Algebra is completed by examining the only type of exponent not yet investigated -- the variable, or placeholder. This is the start of a study of literal-degree relations and is the basis for the development of exponential and logarithmic functions.

It is only after considering all possible degrees, that we can say we have studied a complete course in Algebra. In that context, it is quite artificial to define, for everyone, what "Algebra 1" is, or "Algebra 2," or even "Pre-Algebra." **The logical scope of Algebra covers relations of all degrees, including numeric and literal, while the sequence of concepts begins with a mastery of first-degree relations and grows systematically to include increasingly more sophisticated degrees.**

One more organizational quality is noteworthy here. **The normal flow of each unit is based on the logical introduction of any new mathematical symbolism.** First, the new "thing" is defined and described in detail. Then, operations involving the new "thing" are explored. Finally, relations involving this new "thing" are examined and strategies are developed to resolve them. This cycle is introduced and explained in Unit I and is evidenced in each successive unit. For example, in Unit V, polynomials are introduced. This is new mathematical symbolism for the student, and it must be defined carefully. Then, operations with polynomials must be examined. All of this culminates, of course, in learning to solve relations with polynomials. This l**ogical cycle of exploration in mathematics is helpful to students**, providing them with some anticipation of the levels of exploration necessary to develop algebraic concepts.

Please understand that the organizational argument presented here is not meant to stifle the creativity of the instructor. Neither should it prohibit the instructor from utilizing a modular approach to concept development. It does, however, serve to remedy the fragmented, isolated topic, "chapter" approach to a subject which has been traditionally presented to us in "textbooks" without that element of developmental continuity. To that end, it speaks loudly to the curricular issues which all instructors face and the attitudinal issues students deal with when they are presented with the fact that "everyone must pass Algebra."

Quiz Form A

Name ______________________

Class ________ **Date** ________ **Score** ______

Unit I – The Structure of Mathematics
Part A – Mathematics as a Language

Lesson 1 – Mathematical Parts of Speech

Tell what part of mathematical speech each of the following is.

1. ( ) Grouping Symbol

2. m Placeholder Symbol

3. $\neq$ Relation Symbol

4. $\div$ Operation Symbol

5. $\square$ Placeholder Symbol

6. 88 Number Symbol

7. $\cdot$ Operation Symbol

8. [ ] Grouping Symbol

9. $>$ Relation Symbol

10. 18.7 Number Symbol

Quiz Form B Name ____________________

Class ________ Date ________ Score ________

Unit I – The Structure of Mathematics
Part A – Mathematics as a Language
Lesson 1 – Mathematical Parts of Speech

Tell what part of mathematical speech each of the following is.

1. 101.01 Number Symbol

2. $<$ Relation Symbol

3. $\{\ \}$ Grouping Symbol

4. $\div$ Operation Symbol

5. 13 Number Symbol

6. $\triangle$ Placeholder Symbol

7. x Placeholder Symbol

8. $\not>$ Relation Symbol

9. $-$ Operation Symbol

10. $(\ )$ Grouping Symbol

Quiz Form A

Name ______________________

Class ________ Date ________ Score ______

Unit I – The Structure of Mathematics
Part A – Mathematics as a Language
Lesson 2 – Mathematical Expressions

Tell whether each of the following expressions is an open phrase, closed phrase, open sentence, or closed sentence.

1. $5-3=6\cdot 2$ closed sentence (relation symbol, but no placeholder)

2. $9x+4$ open phrase (placeholder, but no relation symbol)

3. $\sqrt{y+4}$ open phrase (placeholder, but no relation symbol)

4. $x-6=3$ open sentence (placeholder and relation symbol)

5. $6y+2=50$ open sentence (placeholder and relation symbol)

6. $9-6=3$ closed sentence (relation symbol, but no placeholder)

7. $3\cdot 7+6$ closed phrase (no placeholder and no relation symbol)

8. $(9-5)8$ closed phrase (no placeholder and no relation symbol)

Name ______________________

Unit I, Part A, Lesson 2, Quiz Form A
– Continued –

In the following exercises, take the appropriate mathematical action with each expression. If you need a domain, use $\{4, 5, 6\}$. If you need a replacement set, use $\{10, 11, 12\}$.

9. $2x + x$ Domain $\{4, 5, 6\}$ Range $\{12, 15, 18\}$

Using domain $\{4, 5, 6\}$, we substitute and evaluate:

$2(4) + (4) \rightarrow 8 + 4 \rightarrow 12$

$2(5) + (5) \rightarrow 10 + 5 \rightarrow 15$

$2(6) + (6) \rightarrow 12 + 6 \rightarrow 18$

So the range is $\{12, 15, 18\}$

10. $5x + 9 = 59$ $\{10\}$

Substitute from $\{10, 11, 12\}$, and obtain a solution set of values that make the expression true.

$5(10) + 9 = 59$ T

$5(11) + 9 = 59$ F

$5(12) + 9 = 59$ F

So our solution set is $\{10\}$

11. $\frac{13+9}{2} > 5 \cdot 3$ False

Closed sentence, so tell if it is true or false.

$\frac{13+9}{2} > 5 \cdot 3$

$\frac{22}{2} > 15$

$11 > 15$ F

12. $\frac{x}{2} < 13$ $\{10, 11, 12\}$

Substitute from $\{10, 11, 12\}$, and obtain a solution set of values that make the expression true.

$\frac{10}{2} < 13 \rightarrow 5 < 13$ T

$\frac{11}{2} < 13 \rightarrow 5\frac{1}{2} < 13$ T

$\frac{12}{2} < 13 \rightarrow 6 < 13$ T

So our solution set is $\{10, 11, 12\}$.

Quiz Form B

Name ______________________

Class ________ Date ________ Score ______

Unit I – The Structure of Mathematics
Part A – Mathematics as a Language

Lesson 2 – Mathematical Expressions

Tell whether each of the following expressions is an open phrase, closed phrase, open sentence, or closed sentence.

1. $(7+2)\cdot 3$ closed phrase (no placeholder and no relation symbol)

2. $6\cdot 2=7+5$ closed sentence (relation symbol, but no placeholder)

3. $y+4=11$ open sentence (placeholder and relation symbol)

4. $6x+2=14$ open sentence (placeholder and relation symbol)

5. $7-2\cdot 3$ closed phrase (no placeholder and no relation symbol)

6. $2y-8$ open phrase (placeholder, but no relation symbol)

7. $\sqrt{y-3}$ open phrase (placeholder, but no relation symbol)

8. $2+4=7$ closed sentence (relation symbol, but no placeholder)

Name ______________________

Unit I, Part A, Lesson 2, Quiz Form B
– Continued –

In the following exercises, take the appropriate mathematical action with each expression. If you need a domain, use $\{4, 5, 6\}$. If you need a replacement set, use $\{10, 11, 12\}$.

9. $\frac{x}{2} > 3$ $\{10, 11, 12\}$

Substitute from $\{10, 11, 12\}$, and obtain a solution set of values that make the expression true.

$\frac{10}{2} > 3 \rightarrow 5 > 3$ T

$\frac{11}{2} > 3 \rightarrow 5\frac{1}{2} > 3$ T

$\frac{12}{2} > 3 \rightarrow 6 > 3$ T

So our solution set is $\{10, 11, 12\}$

10. $3x - 2$ Domain $\{4, 5, 6\}$ Range $\{10, 13, 16\}$

Using domain $\{4, 5, 6\}$, we substitute and evaluate:

$3(4) - 2 \rightarrow 12 - 2 \rightarrow 10$

$3(5) - 2 \rightarrow 15 - 2 \rightarrow 13$

$3(6) - 2 \rightarrow 18 - 2 \rightarrow 16$

So the range is $\{10, 13, 16\}$

11. $4x - 3 = 41$ $\{11\}$

Substitute from $\{10, 11, 12\}$, and obtain a solution set of values that make the expression true.

$4(10) - 3 = 41 \rightarrow 40 - 3 = 41$ F

$4(11) - 3 = 41 \rightarrow 44 - 3 = 41$ T

$4(12) - 3 = 41 \rightarrow 48 - 3 = 41$ F

So our solution set is $\{11\}$.

12. $6 \div 2 < \frac{10-5}{5}$ False

Closed sentence, so tell if it is true or false.

$6 \div 2 < \frac{10-5}{5}$

$3 < \frac{5}{5}$

$3 < 1$ F

Quiz Form A Name ______________________

Class __________ Date __________ Score ______

Unit I – The Structure of Mathematics
Part A – Mathematics as a Language
Lesson 3 – Translation of Mathematical Symbols

Translate the following English phrases into expressions with mathematical symbols.

1. *x* divided by 7, decreased by 2

   $\frac{x}{7} - 2$ or $(x \div 7) - 2$

2. the sum of *x* and *y*

   $x + y$

3. 15 less than the total of *x* and *y*

   $(x + y) - 15$

4. the difference between *a* and *b*, multiplied by the product of *x* and *y*

   $(x \cdot y)(a - b)$

Translate the following English sentences into expressions with mathematical symbols.

5. The product of 4 and a number is 20.

   $4m = 20$

6. 14 plus, 3 times a number is 59.

   $14 + 3a = 59$

7. 25 is less than the quotient of 9 and *x*.

   $25 < \frac{9}{x}$ or $25 < 9 \div x$

8. The sum of 20 and the product of 9 and *x* is greater than 110.

   $20 + (9 \cdot x) > 110$

Quiz Form B Name ______________________

Class __________ Date ____________ Score ________

Unit I – The Structure of Mathematics
Part A – Mathematics as a Language
Lesson 3 – Translation of Mathematical Symbols

Translate the following English phrases into expressions with mathematical symbols.

1. 12 more than the difference of x and y

$(x-y)+12$

2. the quotient of 7 and 3

$\frac{7}{3}$ or $7 \div 3$

3. 32 greater than the sum of a and b

$(a+b)+32$

4. the sum of m and n, multiplied by the quotient of 1 and 2

$(m+n)\cdot\frac{1}{2}$

Translate the following English sentences into expressions with mathematical symbols.

5. 17 is the difference between 63 and a number.

$17 = 63 - n$

6. 34 is greater than the sum of 7 and y.

$34 > 7+y$

7. 81 minus, 4 times a number is 21.

$81-4x=21$

8. The difference of 10 and a number is less than 52.

$10-n < 52$

Quiz Form A

Name ______________________

Class ________ **Date** ________ **Score** ______

Unit I – The Structure of Mathematics
Part B – Further Investigation of Number Symbols
Lesson 1 – The Development of Our Number System

Match the term in column one with the correct phrase or symbol(s) from column two by placing the correct letter in the blank.

Answer		Term		Column Two
D	1.	Integers	A.	{ }
G	2.	Uniqueness	B.	{0, 1, 2, 3, ...}
H	3.	Real Numbers	C.	must get an answer
K	4.	Rational Numbers	D.	$\{\ldots {}^-3, {}^-2, {}^-1, 0, 1, 2, 3, \ldots\}$
J	5.	Closure	E.	$\{x \mid x \text{ is even}\}$
I	6.	Irrational Numbers	F.	{1, 2, 3, 4, ...}
F	7.	Natural Numbers	G.	get only one answer for each combination of numbers and operations
B	8.	Whole Numbers	H.	set of numbers consisting of both rational and irrational numbers
C	9.	Existence	I.	numbers that cannot be represented as a ratio of integers
E	10.	Set Builder Notation	J.	answers must be in the set we are working in
			K.	numbers that can be represented as a ratio $\frac{a}{b}$ where a and b are integers and $b \neq 0$.

Name ______________________

Unit I, Part B, Lesson 1, Quiz Form A
– Continued –

11. In each of the following sets, tell if **a.** multiplication, **b.** subtraction, **c.** division, and **d.** addition are well-defined operations. If not, give an example to show why not.

$\{0, 1, 2, 3, 4, ...\}$

a. multiplication – yes, multiplying whole numbers gives whole numbers.

b. subtraction – no Examples will vary. $2-5 = {}^{-}3$, and ${}^{-}3$ is not a whole number.

c. division – no Examples will vary. $2 \div 5 = \frac{2}{5}$, and $\frac{2}{5}$ is not a whole number.

d. addition – yes, adding whole numbers gives whole numbers.

$\{1, 3, 5, 7, ...\}$

a. multiplication – yes, multiplying positive odd integers gives positive odd integers

b. subtraction – no Examples will vary. $1-3 = {}^{-}2$, and ${}^{-}2$ is not in the set.

c. division – no Examples will vary. $5 \div 7 = \frac{5}{7}$, and $\frac{5}{7}$ is not in the set.

d. addition – no Examples will vary. $3+5=8$, and 8 is not in the set.

Quiz Form B Name ______________________

Class __________ Date __________ Score ______

Unit I – The Structure of Mathematics
Part B – Further Investigation of Number Symbols
Lesson 1 – The Development of Our Number System

Match the term in column one with the correct phrase or symbol(s) from column two by placing the correct letter in the blank.

Answer	Term		Choice
J	**1.** Closure	**A.**	$\{\ \}$
F	**2.** Natural Numbers	**B.**	$\{0, 1, 2, 3, ...\}$
C	**3.** Existence	**C.**	must get an answer
E	**4.** Set Builder Notation	**D.**	$\{... {}^{-}3, {}^{-}2, {}^{-}1, 0, 1, 2, 3, ...\}$
K	**5.** Rational Numbers	**E.**	$\{x \mid x \text{ is even}\}$
H	**6.** Real Numbers	**F.**	$\{1, 2, 3, 4, ...\}$
G	**7.** Uniqueness	**G.**	get only one answer for each combination of numbers and generations
D	**8.** Integers	**H.**	set of numbers consisting of both rational and irrational numbers
B	**9.** Whole Numbers	**I.**	numbers that cannot be represented as a ratio of integers
I	**10.** Irrational Numbers	**J.**	answers must be in the set we are working in
		K.	numbers that can be represented as a ratio $\frac{a}{b}$ where a and b are integers and $b \neq 0$.

Name ______________________

Unit I, Part B, Lesson 1, Quiz Form B
– Continued –

11. In each of the following sets, tell if **a.** addition, **b.** division, **c.** multiplication, and **d.** subtraction are well-defined operations. If not, give an example to show why not.

$$\{0, 2, 4, 6, 8, ...\}$$

a. addition – yes, adding even whole numbers gives even whole numbers.

b. division – no Answers will vary. $4 \div 6 = \frac{4}{6}$, and $\frac{4}{6}$ is not an even whole number.

c. multiplication – yes, multiplying even whole numbers gives even whole numbers.

d. subtraction – no Answers will vary. $2 - 6 = {}^-4$, and ${}^-4$ is not an even whole number.

$$\{1, 2, 3, 4, ...\}$$

a. addition – yes, adding natural numbers gives natural numbers.

b. division – no Answers will vary. $4 \div 3 = \frac{4}{3}$, and $\frac{4}{3}$ is not a natural number.

c. multiplication – yes, multiplying natural numbers gives natural numbers.

d. subtraction – no Answers will vary. $1 - 5 = {}^-4$, and ${}^-4$ is not a natural number.

Quiz Form A

Name ____________________

Class ________ Date __________ Score ______

Unit I – The Structure of Mathematics
Part B – Further Investigation of Number Symbols
Lesson 2 – Fraction Forms and Decimal Forms
Lesson 3 – Changing Fraction Forms to Decimal Forms

Complete the equivalent form indicated for each of the following fractions.

1. $\frac{7}{9} = \frac{35}{45}$ $\quad \frac{7}{9} \cdot \frac{5}{5} = \frac{35}{45}$

2. $\frac{11}{5} = \frac{77}{35}$ $\quad \frac{11}{5} \cdot \frac{7}{7} = \frac{77}{35}$

Simplify each of the following fractions by rewriting them in factored form and finding all the "ones" you can.

3. $\frac{42}{60} = \frac{2 \cdot 3 \cdot 7}{2 \cdot 2 \cdot 3 \cdot 5} = \frac{7}{10}$

4. $\frac{16}{80} = \frac{2 \cdot 2 \cdot 2 \cdot 2}{2 \cdot 2 \cdot 2 \cdot 2 \cdot 5} = \frac{1}{5}$

5. $\frac{51}{17} = \frac{3 \cdot 17}{17} = 3$

6. $\frac{7}{33} = \frac{7}{33}$

Name ______________________________

Unit I, Part B, Lessons 2 and 3, Quiz Form A
– Continued –

Find a terminating decimal, equivalent to each of the following fractions. Show your work.

7. $\frac{5}{4}$ = 1.25

$$\begin{array}{r} 1.25 \\ 4\overline{)5.00} \\ \underline{4} \\ 1\,0 \\ \underline{8} \\ 20 \\ \underline{20} \\ 0 \end{array}$$

8. $\frac{3}{8}$ = .375

$$\begin{array}{r} .375 \\ 8\overline{)3.000} \\ \underline{2\,4} \\ 60 \\ \underline{56} \\ 40 \\ \underline{40} \\ 0 \end{array}$$

Find a repeating decimal, equivalent to each of the following fractions. Show your work.

9. $\frac{5}{11}$ = $0.\overline{45}$

$$\begin{array}{r} .4545\ldots \\ 11\overline{)5.0000} \\ \underline{4\,4} \\ 60 \\ \underline{55} \\ 50 \\ \underline{44} \\ 60 \\ \underline{55} \\ 5 \end{array}$$

This process will continue forever, so we write $0.\overline{45}$.

10. $\frac{5}{7}$ = $0.\overline{714285}$

$$\begin{array}{r} .71428571\ldots \\ 7\overline{)5.00000000} \\ \underline{4\,9} \\ 10 \\ \underline{7} \\ 30 \\ \underline{28} \\ 20 \\ \underline{14} \\ 60 \\ \underline{56} \\ 40 \\ \underline{35} \\ 50 \\ \underline{49} \\ 10 \\ \underline{7} \\ 3 \end{array}$$

This process will continue forever, so we write $0.\overline{714285}$

Quiz Form B

Name ____________________

Class ________ Date ________ Score ______

Unit I – The Structure of Mathematics
Part B – Further Investigation of Number Symbols
Lesson 2 – Fraction Forms and Decimal Forms
Lesson 3 – Changing Fraction Forms to Decimal Forms

Complete the equivalent form indicated for each of the following fractions.

1. $\frac{5}{13} = \frac{30}{78}$ $\quad \frac{5}{13} \cdot \frac{6}{6} = \frac{30}{78}$

2. $\frac{5}{8} = \frac{20}{32}$ $\quad \frac{5}{8} \cdot \frac{4}{4} = \frac{20}{32}$

Simplify each of the following fractions by rewriting them in factored form and finding all the "ones" you can.

3. $\frac{18}{72} = \frac{2 \cdot 3 \cdot 3}{2 \cdot 2 \cdot 2 \cdot 3 \cdot 3} = \frac{1}{4}$

4. $\frac{13}{55} = \frac{13}{55}$

5. $\frac{55}{121} = \frac{5 \cdot 11}{11 \cdot 11} = \frac{5}{11}$

6. $\frac{15}{90} = \frac{3 \cdot 5}{2 \cdot 3 \cdot 3 \cdot 5} = \frac{1}{6}$

Name ______________________________

Unit I, Part B, Lessons 2 and 3, Quiz Form B
– Continued –

Find a terminating decimal, equivalent to each of the following fractions. Show your work.

7. $\frac{6}{5} = 1.2$

$$
\begin{array}{r}
1.2 \\
5\overline{)6.0} \\
\underline{5} \\
1\,0 \\
\underline{1\,0} \\
0
\end{array}
$$

8. $\frac{5}{16} = 0.3125$

$$
\begin{array}{r}
.3125 \\
16\overline{)5.0000} \\
\underline{4\,8} \\
20 \\
\underline{16} \\
40 \\
\underline{32} \\
80 \\
\underline{80} \\
0
\end{array}
$$

Find a repeating decimal, equivalent to each of the following fractions. Show your work.

9. $\frac{7}{9} = 0.\overline{7}$

$$
\begin{array}{r}
.777\ldots \\
9\overline{)7.000} \\
\underline{6\,3} \\
70 \\
\underline{63} \\
70 \\
\underline{63} \\
7
\end{array}
$$

This process will continue forever, so we write $0.\overline{7}$.

10. $\frac{8}{13} = 0.\overline{615384}$

$$
\begin{array}{r}
.61538461\ldots \\
13\overline{)8.00000000} \\
\underline{7\,8} \\
20 \\
\underline{13} \\
70 \\
\underline{65} \\
50 \\
\underline{39} \\
110 \\
\underline{104} \\
60 \\
\underline{52} \\
80 \\
\underline{78} \\
20 \\
\underline{13} \\
7
\end{array}
$$

This process will continue forever, so we write $0.\overline{615384}$.

Quiz Form A

Name ______________________

Class ________ Date __________ Score ______

Unit I – The Structure of Mathematics
Part B – Further Investigation of Number Symbols

Lesson 4 – Changing Decimal Forms to Fraction Forms
Lesson 5 – Percent

Find a fraction, equivalent to each of the following terminating decimal numbers.

1. .12

$.12 = \frac{12}{100} = \frac{3}{25}$

2. 7.05

$7.05 = \frac{705}{100} = 7\frac{5}{100} = 7\frac{1}{20}$

3. .94101

$\frac{94{,}101}{100{,}000}$

Find a fraction, equivalent to the following repeating decimal numbers. Show your work.

4. $.\overline{23}$

$100f = 23.\overline{23}$
$-1f = .\overline{23}$
$99f = 23$

$f = \frac{23}{99}$

$.\overline{23} = \frac{23}{99}$

5. $.\overline{3}$

$10f = 3.\overline{3}$
$-1f = .\overline{3}$
$9f = 3$

$f = \frac{3}{9}$ or $\frac{1}{3}$

$.\overline{3} = \frac{1}{3}$

6. $.\overline{234}$

$1000f = 234.\overline{234}$
$-1f = .\overline{234}$
$999f = 234$

$f = \frac{234}{999}$

$.\overline{234} = \frac{234}{999}$ or $\frac{78}{333}$ or $\frac{26}{111}$

For each of the following fractions, find **a.** an equivalent decimal form and **b.** an equivalent percent form.

7. $\frac{13}{100}$ **a.** .13 **b.** 13%

a.) $13.00 \div 100 = .13$ (100 → 300 → 300 → 0)

b.) To change .13 to an equivalent percent form, move the decimal point two places to the right, and attach a % sign.

8. $\frac{13}{25}$ **a.** .52 **b.** 52%

a.) $13.00 \div 25 = .52$ (125 → 50 → 50 → 0)

b.) To change .52 to an equivalent percent form, move the decimal point two places to the right, and attach a % sign.

Name ________________________________

Unit I, Part B, Lessons 4 and 5, Quiz Form A
– Continued –

For each of the following decimal numbers, find **a.** an equivalent fraction form and **b.** an equivalent percent form.

9. .46 **a.** $\frac{46}{100}$ or $\frac{23}{50}$ **b.** 46%

a.) .46 is read "46 hundredths", and this is represented by $\frac{46}{100}$.

b.) To change .46 to an equivalent percent form, move the decimal point two places to the right, and attach a % sign.

10. .0021 **a.** $\frac{21}{10{,}000}$ **b.** .21%

a.) .0021 is read "21 ten thousandths", and is represented by $\frac{21}{10{,}000}$.

b.) To change .0021 to an equivalent percent form, move the decimal point two places to the right and attach a % sign.

For each of the following percents, find **a.** an equivalent fraction form and **b.** an equivalent decimal form.

11. 43% **a.** $\frac{43}{100}$ **b.** .43

a.) Since percent is a comparison to 100, we can write 43% as $\frac{43}{100}$.

b.) Reading $\frac{43}{100}$ properly gives us "43 hundredths" which is .43.

12. 4.2% **a.** $\frac{42}{1{,}000}$ **b.** .042

a.) Since percent is a comparison to 100, we can write $\frac{4.2}{100}$. Since the numerator is not a whole number, multiply by $\frac{10}{10}$.

$$\frac{4.2}{100} \times \frac{10}{10} = \frac{42}{1{,}000}$$

b.) Reading $\frac{42}{1{,}000}$ properly gives us "forty-two thousandths" which is .042.

Quiz Form B

Name ______________________

Class ________ Date __________ Score ______

Unit I – The Structure of Mathematics
Part B – Further Investigation of Number Symbols

Lesson 4 – Changing Decimal Forms to Fraction Forms
Lesson 5 – Percent

Find a fraction, equivalent to each of the following terminating decimal numbers.

1. .84

$.84 = \frac{84}{100} = \frac{21}{25}$

2. 11.12

$11.12 = \frac{1{,}112}{100} = \frac{278}{25}$

3. .16707

$.16707 = \frac{16{,}707}{100{,}000}$

Find a fraction, equivalent to the following repeating decimal numbers. Show your work.

4. $.\overline{6}$

$10f = 6.\overline{6}$
$-1f = .\overline{6}$
$9f = 6$

$f = \frac{6}{9}$ or $\frac{2}{3}$

$.\overline{6} = \frac{2}{3}$

5. $.\overline{47}$

$100f = 47.\overline{47}$
$-\ f = .\overline{47}$
$99f = 47$

$f = \frac{47}{99}$

$.\overline{47} = \frac{47}{99}$

6. $.\overline{781}$

$1{,}000f = 781.\overline{781}$
$-1f = .\overline{781}$
$999f = 781$

$f = \frac{781}{999}$

$.\overline{781} = \frac{781}{999}$

For each of the following fractions, find **a.** an equivalent decimal form and **b.** an equivalent percent form.

7. $\frac{87}{100}$ **a.** .87 **b.** 87%

a.) $87.00 \div 100 = .87$
800
700
700
0

b.) To change .87 to an equivalent percent form, move the decimal point two places to the right, and attach a percent sign.

8. $\frac{17}{50}$ **a.** .34 **b.** 34%

a.) $17.00 \div 50 = .34$
150
200
200
0

b.) To change .34 to an equivalent percent form, move the decimal point two places to the right, and attach a percent sign.

Name ______________________________

Unit I, Part B, Lessons 4 and 5, Quiz Form B
– Continued –

For each of the following decimal numbers, find **a.** an equivalent fraction form and **b.** an equivalent percent form.

9. .84 **a.** $\frac{84}{100}$ or $\frac{21}{25}$ **b.** 84%

a.) .84 is read "eighty-four hundredths," and this is represented by $\frac{84}{100}$ or $\frac{21}{25}$.

b.) To change .84 to an equivalent percent form, move the decimal point two places to the right, and attach a percent sign.

10. .073 **a.** $\frac{73}{1,000}$ **b.** 7.3%

a.) .073 is read "seventy-three thousandths", and this is represented by $\frac{73}{1,000}$.

b.) To change .073 to an equivalent percent form, move the decimal point two places to the right, and attach a percent sign.

For each of the following percents, find **a.** an equivalent fraction form and **b.** an equivalent decimal form.

11. 67% **a.** $\frac{67}{100}$ **b.** .67

a.) Since percent is a comparison to 100, we can write 67% as $\frac{67}{100}$.

b.) Reading $\frac{67}{100}$ properly gives us "67 hundredths" which is .67.

12. 16.8% **a.** $\frac{168}{1,000}$ or $\frac{21}{125}$ **b.** .168

a.) Since percent is a comparison to 100, we can write 16.8% as $\frac{16.8}{100}$. Since the numerator is not a whole number, multiply by $\frac{10}{10}$.

$$\frac{16.8}{100} \times \frac{10}{10} = \frac{168}{1,000}$$

$$\frac{168}{1,000} = \frac{\cancel{8} \cdot 21}{\cancel{8} \cdot 125} = \frac{21}{125}$$

b.) Reading $\frac{168}{1,000}$ properly gives us "168 thousandths" which is .168.

Quiz Form A Name ______________________

Class ________ Date __________ Score ______

Unit I – The Structure of Mathematics
Part B – Further Investigation of Number Symbols
Lesson 6 – Primes, Composites, and Factoring
Lesson 7 – Least Common Multiple
Lesson 8 – Greatest Common Factor

Find all factored forms and the prime factored form (PFF).

1. 40 2x20 4x10 5x8 1x40 ____ PFF 2x2x2x5

2. 39 3x13 1x39 ____ ____ ____ PFF 3x13

3. 36 2x18 3x12 4x9 6x6 1x36 PFF 2x2x3x3

Find the Least Common Multiple (LCM) for each of the following, using the method indicated.

4. By listing the multiples of each LCM 56

 7 7 14 21 28 35 42 49 56 63

 8 8 16 24 32 40 48 56 64

5. By factoring each number as a product of primes LCM 2x3x3x5 = 90

 18 2x3x3

 45 3x3x5

Name ______________________________

Unit I, Part B, Lessons 6, 7, and 8, Quiz Form A
– Continued –

Find the Greatest Common Factor (GCF) for each of the following, using the method indicated.

6. By listing the factors of each GCF 5

20 1 2 4 (5) 10 20

15 1 3 (5) 15

7. By factoring each number as a product of primes GCF 2x3 = 6

36 2 x 2 x 3 x 3

42 2 x 3 x 7

Quiz Form B Name ____________________

Class ________ Date __________ Score ______

Unit I – The Structure of Mathematics
Part B – Further Investigation of Number Symbols
Lesson 6 – Primes, Composites, and Factoring
Lesson 7 – Least Common Multiple
Lesson 8 – Greatest Common Factor

Find all factored forms and the prime factored form (PFF).

1. 32 2x16 4x8 1x32 _____ _____ PFF 2x2x2x2x2

2. 27 3x9 1x27 _____ _____ _____ PFF 3x3x3

3. 81 3x27 9x9 1x81 _____ _____ PFF 3x3x3x3

Find the Least Common Multiple (LCM) for each of the following, using the method indicated.

4. By listing the multiples of each LCM 33

11 11 22 (33) 44

3 3 6 9 12 15 18 21 24 27 30 (33) 36

5. By factoring each number as a product of primes LCM 2x2x5x5 = 100

20 2x2x5

25 5x5

Name ______________________________

Unit I, Part B, Lessons 6,7, and 8, Quiz Form B
– Continued –

Find the Greatest Common Factor (GCF) for each of the following, using the method indicated.

6. By listing the factors of each GCF 9

18 1 2 3 6 (9) 18

45 1 3 5 (9) 15 45

7. By factoring each number as a product of primes GCF 2x2x2 = 8

72 2x2x2x3x3

40 2x2x2 x 5

Quiz Form A

Name ______________________

Class ________ Date __________ Score ______

Unit I – The Structure of Mathematics
Part C – Further Investigation of Operation Symbols
Lesson 1 – Order of Operations
Lesson 2 – Properties of Operations
Lesson 3 – Properties of Operations with Special Numbers

Evaluate the following closed phrases.

1. $10+14\div7+3$ ______15______

10 + 2 + 3
12 + 3
15

2. $30-3(7)+4$ ______13______

30 - 21 + 4
9 + 4
13

3. $15\div3+2(7)$ ______19______

5 + 2(7)
5 + 14
19

4. $48\div6-36\div9$ ______4______

8 - 36 ÷ 9
8 - 4
4

5. $36\div12\cdot3+4\cdot2$ ______17______

3 · 3 + 4 · 2
9 + 4 · 2
9 + 8
17

Name ______________________

Unit I, Part C, Lessons 1,2, and 3, Quiz Form A
– Continued –

Identify completely the property of operations illustrated by each of the following.

6. $9 \cdot 7 = 7 \cdot 9$ — Commutative Property of Multiplication

7. $8 + (7 + 5) = (8 + 7) + 5$ — Associative Property of Addition

8. $x \cdot (a + b) = (x \cdot a) + (x \cdot b)$ — Distributive Property of Multiplication over Addition

9. $4 \cdot (3 \cdot 2) = (4 \cdot 3) \cdot 2$ — Associative Property of Multiplication

10. $(x + y) \cdot z = (x \cdot z) + (y \cdot z)$ — Distributive Property of Multiplication over Addition

Identify completely which special property concerning zero or one is used to justify each step in the following problem.

$[(5-5) + 4 \cdot 1] \cdot \frac{1}{4} = n$

11. $[\ 0 \ + 4 \cdot 1] \cdot \frac{1}{4} = n$ — Additive Inverse Property

12. $[\ 0 \ + \ 4\ ] \cdot \frac{1}{4} = n$ — Identity Property of Multiplication

13. $[\ \ 4\ \ ] \cdot \frac{1}{4} = n$ — Identity Property of Addition

14. $1 = n$ — Multiplicative Inverse Property

Quiz Form B

Name ______________________________

Class __________ Date __________ Score ______

Unit I – The Structure of Mathematics
Part C – Further Investigation of Operation Symbols

Lesson 1 – Order of Operations
Lesson 2 – Properties of Operations
Lesson 3 – Properties of Operations with Special Numbers

Evaluate the following closed phrases.

1. $8+14\div2+3$ 18

8 + 7 + 3
15 + 3
18

2. $3(7)+24\div8$ 24

21 + 24 ÷ 8
21 + 3
24

3. $12\div2+3(4)$ 18

6 + 3(4)
6 + 12
18

4. $24\div6-10\div5$ 2

4 − 10 ÷ 5
4 − 2
2

5. $40-5\cdot7+10\div5$ 7

40 − 35 + 10 ÷ 5
40 − 35 + 2
5 + 2
7

Name ______________________________

Unit I, Part C, Lessons 1,2, and 3, Quiz Form B
– Continued –

Identify completely the property of operations illustrated by each of the following.

6. $9+(x+y)=(9+x)+y$ Associative Property of Addition

7. $8\cdot(3+9)=(8\cdot3)+(8\cdot9)$ Distributive Property of Multiplication over Addition

8. $4\cdot8=8\cdot4$ Commutative Property of Multiplication

9. $6\cdot(5\cdot4)=(6\cdot5)\cdot4$ Associative Property of Multiplication

10. $7\cdot(5-4)=(7\cdot5)-(7\cdot4)$ Distributive Property of Multiplication over Subtraction

Identify completely which special property concerning zero or one is used to justify each step in the following problem.

$$\left(5\cdot\frac{1}{5}\right)+[7\cdot1+(3-3)]=n$$

11. $1+[7\cdot1+(3-3)]=n$ Multiplicative Inverse Property

12. $1+[7+(3-3)]=n$ Identity Property of Multiplication

13. $1+[7+0]=n$ Additive Inverse Property

14. $1+[7]=n$ Identity Property of Addition

$8=n$ **Arithmetic Fact**

Quiz Form A Name ______________________

Class ________ Date __________ Score ______

Unit I – The Structure of Mathematics
Part C – Further Investigation of Operation Symbols

Lesson 4 – Operations with Fractions – Multiplication

Lesson 5 – Operations with Fractions – Addition and Subtraction

Lesson 6 – Operations with Fractions – Division

Use the given rectangle to illustrate the multiplication of the two fractions.

1. $\frac{3}{4} \cdot \frac{1}{5}$

$\frac{3}{20}$

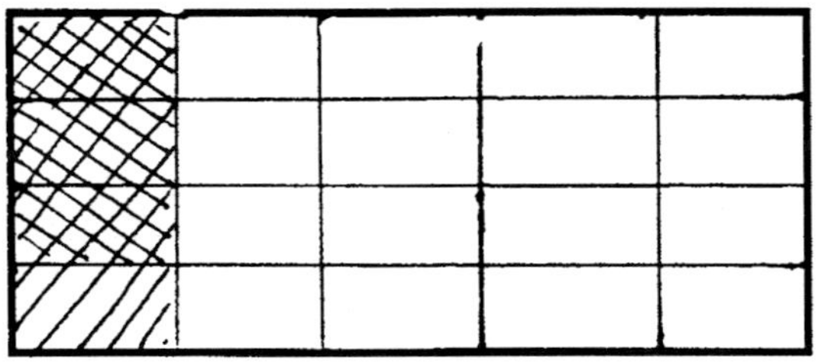

Find the answer to each of the following problems, showing the process of the operation and simplifying your result (if possible). Circle your answer.

2. $\frac{2}{3} \cdot \frac{5}{7} = \frac{2 \times 5}{3 \times 7} = \frac{10}{21}$

3. $\frac{5}{8} \cdot \frac{6}{7} = \frac{5 \times 6}{8 \times 7} = \frac{5 \times 2 \times 3}{2 \times 2 \times 2 \times 7} = \frac{15}{28}$

4. $\frac{12}{35} \cdot \frac{14}{15} = \frac{12 \cdot 14}{35 \cdot 15} = \frac{2 \cdot 2 \cdot 3 \cdot 2 \cdot 7}{5 \cdot 7 \cdot 3 \cdot 5} = \frac{8}{25}$

5. $\frac{15}{9} \cdot \frac{3}{20} = \frac{15 \cdot 3}{9 \cdot 20} = \frac{3 \cdot 5 \cdot 3}{3 \cdot 3 \cdot 2 \cdot 2 \cdot 5} = \frac{1}{4}$

Find the answer to each of the following problems, showing the process of the operation and simplifying your result (if possible). Circle your answer.

6. $\frac{3}{4} + \frac{2}{3} =$

$4 \to 2 \cdot 2$
$3 \to 3$
LCM is $2 \cdot 2 \cdot 3 = 12$

$\frac{3}{4} \cdot 1 + \frac{2}{3} \cdot 1$

$\frac{3}{4} \cdot \frac{3}{3} + \frac{2}{3} \cdot \frac{4}{4}$

$\frac{9}{12} + \frac{8}{12}$

$\frac{17}{12}$

7. $\frac{7}{11} - \frac{1}{3} =$

$11 \to 11$
$3 \to 3$
LCM is $3 \cdot 11 = 33$

$\frac{7}{11} \cdot 1 - \frac{1}{3} \cdot 1$

$\frac{7}{11} \cdot \frac{3}{3} - \frac{1}{3} \cdot \frac{11}{11}$

$\frac{21}{33} - \frac{11}{33}$

$\frac{10}{33}$

Name ______________________________

Unit I, Part C, Lessons 4, 5 and 6, Quiz Form A
– Continued –

$8 \rightarrow 2 \cdot 2 \cdot 2$
$6 \rightarrow 2 \cdot 3$
LCM is $2 \cdot 2 \cdot 2 \cdot 3 = 24$

8. $\frac{2}{9} + \frac{5}{9} =$

$\frac{2+5}{9}$

$\frac{7}{9}$

9. $\frac{15}{8} - \frac{5}{6} =$

$\frac{15}{8} \cdot 1 - \frac{5}{6} \cdot 1$

$\frac{15}{8} \cdot \frac{3}{3} - \frac{5}{6} \cdot \frac{4}{4}$

$\frac{45}{24} - \frac{20}{24} = \frac{25}{24}$

Evaluate each of the following closed phrases by remembering that division means fractions. Be sure to simplify your answer, then circle it.

10. $(5 \div 8) \div (3 \div 4)$

$\frac{5}{8} \div \frac{3}{4}$

$\frac{5}{8} \cdot \frac{4}{3}$

$\frac{5 \cdot 4}{8 \cdot 3} \rightarrow \frac{5 \cdot 2 \cdot 2}{2 \cdot 2 \cdot 2 \cdot 3} = \frac{5}{6}$

11. $(9 \div 10) \div (2 \div 5)$

$\frac{9}{10} \div \frac{2}{5}$

$\frac{9}{10} \cdot \frac{5}{2}$

$\frac{9 \cdot 5}{10 \cdot 2} \rightarrow \frac{3 \cdot 3 \cdot 5}{2 \cdot 5 \cdot 2} = \frac{9}{4}$

Evaluate each of the following by performing the indicated division, remembering the pattern developed in the lesson. Circle your answer.

12. $\frac{5}{6} \div \frac{9}{10}$

$\frac{5}{6} \cdot \frac{10}{9}$

$\frac{5 \cdot 10}{6 \cdot 9} \rightarrow \frac{5 \cdot 5 \cdot 2}{2 \cdot 3 \cdot 3 \cdot 3} = \frac{25}{27}$

13. $\frac{2}{5} \div \frac{4}{7}$

$\frac{2}{5} \cdot \frac{7}{4}$

$\frac{2 \cdot 7}{5 \cdot 4} \rightarrow \frac{2 \cdot 7}{5 \cdot 2 \cdot 2} = \frac{7}{10}$

14. $\frac{3}{8} \div \frac{5}{11}$

$\frac{3}{8} \cdot \frac{11}{5}$

$\frac{3 \cdot 11}{8 \cdot 5} = \frac{33}{40}$

15. $\frac{4}{9} \div \frac{16}{27}$

$\frac{4}{9} \cdot \frac{27}{16}$

$\frac{4 \cdot 27}{9 \cdot 16}$

$\frac{2 \cdot 2 \cdot 3 \cdot 3 \cdot 3}{3 \cdot 3 \cdot 2 \cdot 2 \cdot 2 \cdot 2} = \frac{3}{4}$

Looking back...

16. Simplify: $16 \cdot 30 - (9 + 4)$

$16 \cdot 30 - 13$
$480 - 13$
467

17. $(8+7)+4 = 8+(7+4)$ is an illustration of which property of operations? (Identify completely.)

Associative Property of Addition, since the numbers have been grouped (associated) differently.

Quiz Form B

Name ______________________

Class ________ Date __________ Score ______

Unit I – The Structure of Mathematics
Part C – Further Investigation of Operation Symbols

Lesson 4 – Operations with Fractions – Multiplication
Lesson 5 – Operations with Fractions – Addition and Subtraction
Lesson 6 – Operations with Fractions – Division

Use the given rectangle to illustrate the multiplication of the two fractions.

1. $\frac{3}{5} \cdot \frac{1}{4}$

$\frac{3}{20}$

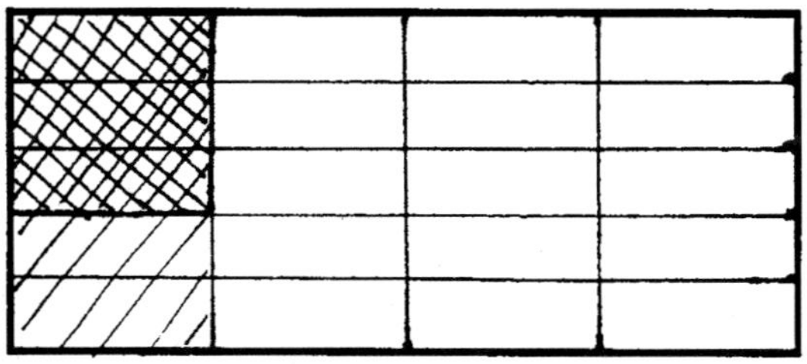

Find the answer to each of the following problems, showing the process of the operation and simplifying your result (if possible). Circle your answer.

2. $\frac{2}{5} \cdot \frac{7}{11} = \frac{2 \cdot 7}{5 \cdot 11} = \frac{14}{55}$

3. $\frac{5}{9} \cdot \frac{3}{7} = \frac{5 \cdot 3}{9 \cdot 7} = \frac{5 \cdot 3}{3 \cdot 3 \cdot 7} = \frac{5}{21}$

4. $\frac{14}{45} \cdot \frac{27}{56} = \frac{14 \cdot 27}{45 \cdot 56} = \frac{2 \cdot 7 \cdot 3 \cdot 3 \cdot 3}{3 \cdot 3 \cdot 5 \cdot 2 \cdot 2 \cdot 2 \cdot 7} = \frac{3}{20}$

5. $\frac{17}{11} \cdot \frac{22}{51} = \frac{17 \cdot 22}{11 \cdot 51} = \frac{17 \cdot 2 \cdot 11}{11 \cdot 3 \cdot 17} = \frac{2}{3}$

Find the answer to each of the following problems, showing the process of the operation and simplifying your result (if possible). Circle your answer.

6. $\frac{5}{8} + \frac{4}{5} =$

$8 \rightarrow 2 \cdot 2 \cdot 2$
$5 \rightarrow 5$
LCM is $2 \cdot 2 \cdot 2 \cdot 5 = 40$

$\frac{5}{8} \cdot 1 + \frac{4}{5} \cdot 1$

$\frac{5}{8} \cdot \frac{5}{5} + \frac{4}{5} \cdot \frac{8}{8}$

$\frac{25}{40} + \frac{32}{40}$

$\frac{57}{40}$

7. $\frac{3}{7} - \frac{1}{5} =$

$7 \rightarrow 7$
$5 \rightarrow 5$
LCM is $7 \cdot 5 = 35$

$\frac{3}{7} \cdot 1 - \frac{1}{5} \cdot 1$

$\frac{3}{7} \cdot \frac{5}{5} - \frac{1}{5} \cdot \frac{7}{7}$

$\frac{15}{35} - \frac{7}{35}$

$\frac{8}{35}$

Name ______________________________

Unit I, Part C, Lessons 4, 5 and 6, Quiz Form B
– Continued –

8. $\frac{3}{11} + \frac{7}{11} =$

$\frac{3+7}{11}$

$\frac{10}{11}$

9. $\frac{17}{9} - \frac{5}{6} =$

$9 \rightarrow 3 \cdot 3$
$6 \rightarrow 2 \cdot 3$
LCM is $2 \cdot 3 \cdot 3 = 18$

$\frac{17}{9} \cdot 1 - \frac{5}{6} \cdot 1$

$\frac{17}{9} \cdot \frac{2}{2} - \frac{5}{6} \cdot \frac{3}{3}$

$\frac{34}{18} - \frac{15}{18}$

$\frac{19}{18}$

Evaluate each of the following closed phrases by remembering that division means fractions. Be sure to simplify your answer, then circle it.

10. $(7 \div 9) \div (2 \div 3)$

$\frac{7}{9} \div \frac{2}{3}$

$\frac{\frac{7}{9}}{\frac{2}{3}} \rightarrow \frac{\frac{7}{9} \cdot \frac{3}{2}}{\frac{2}{3} \cdot \frac{3}{2}} = \frac{\frac{7 \cdot 3}{3 \cdot 3 \cdot 2}}{1} = \frac{\frac{7}{6}}{1} = \frac{7}{6}$

11. $(11 \div 12) \div (3 \div 4)$

$\frac{11}{12} \div \frac{3}{4}$

$\frac{\frac{11}{12}}{\frac{3}{4}} \rightarrow \frac{\frac{11}{12} \cdot \frac{4}{3}}{\frac{3}{4} \cdot \frac{4}{3}} \rightarrow \frac{\frac{11 \cdot 2 \cdot 2}{2 \cdot 2 \cdot 3 \cdot 3}}{1} \rightarrow \frac{\frac{11}{9}}{1} \rightarrow \frac{11}{9}$

Evaluate each of the following by performing the indicated division, remembering the pattern developed in the lesson. Circle your answer.

12. $\frac{3}{7} \div \frac{14}{15}$

$\frac{3}{7} \cdot \frac{15}{14}$

$\frac{3 \cdot 15}{7 \cdot 14} \rightarrow \frac{3 \cdot 3 \cdot 5}{7 \cdot 2 \cdot 7} = \frac{45}{98}$

13. $\frac{5}{7} \div \frac{10}{9}$

$\frac{5}{7} \cdot \frac{9}{10}$

$\frac{5 \cdot 9}{7 \cdot 10} \rightarrow \frac{5 \cdot 3 \cdot 3}{7 \cdot 2 \cdot 5} = \frac{9}{14}$

14. $\frac{5}{8} \div \frac{7}{11}$

$\frac{5}{8} \cdot \frac{11}{7}$

$\frac{5 \cdot 11}{8 \cdot 7} \rightarrow \frac{55}{56}$

15. $\frac{5}{13} \div \frac{20}{39}$

$\frac{5}{13} \cdot \frac{39}{20}$

$\frac{5 \cdot 39}{13 \cdot 20}$

$\frac{5 \cdot 3 \cdot 13}{13 \cdot 2 \cdot 2 \cdot 5} = \frac{3}{4}$

Looking back...

16. Simplify: $(8-2) \cdot 4 + 7$

$6 \cdot 4 + 7$

$24 + 7$

31

17. $6 \cdot (4-c) = (6 \cdot 4) - (6 \cdot c)$ is an illustration of which property of operations? (Identify completely.)

Distributive Property of Multiplication over Subtraction, since the multiplication by 6 has been handed out (distributed) to both the 4 and c.

Quiz Form A

Name ______________________

Class ________ Date ________ Score ______

Unit I – The Structure of Mathematics
Part C – Further Investigations of Operation Symbols
Lesson 7 – Operations with Decimals

Evaluate each of the following by performing the indicated operation.

1. 5.371 + 7.35   12.721

```
 5.371
 7.35
------
12.721
```

2. 15.79 – 8.53   7.26

```
15.79
 8.53
-----
 7.26
```

3. 17.2 – 9.86   7.34

```
 17.20
- 9.86
------
  7.34
```

4. 32.98 · 4.3   141.814

```
  32.98  -> 2 places to the right
    4.3  -> 1 place to the right
  -----
   9894
 13192
-------
141.814  -> total of 3 places to the right
```

Name ______________________________

Unit I, Part C, Lesson 7, Quiz Form A
– Continued –

5. $2.1 \div .05$ 42

$$.05\overline{)2.10} = 42.$$
20
10
10
0

6. $27.48 \div 4$ 6.87

$$4\overline{)27.48} = 6.87$$
24
34
32
28
28
0

7. $13.3 \cdot .005$ .0665

13.3 → 1 place to the right
.005 → 3 places to the right
.0665 → total of 4 places to the right

8. $10.608 \div 4.42$ 2.4

$$4.42\overline{)10.608} = 2.4$$
884
1768
1768
0

Quiz Form B

Name ______________________

Class __________ Date __________ Score ______

Unit I – The Structure of Mathematics
Part C – Further Investigation of Operation Symbols

Lesson 7 – Operations with Decimals

Evaluate each of the following by performing the indicated operation.

1. .8 · 7.3   5.84

7.3 → 1 place to the right
.8 → 1 place to the right
5.84 → total of 2 places to the right

2. .8149 + 6.74   7.5549

.8149
6.74
7.5549

3. .94882 ÷ 6.6 $.1437\overline{60}$

6.6).94882 0000 = .14376060
66
288
264
248
198
502
462
400
396
400
396
40

We can see that this process will continue, so we write $.1437\overline{60}$

4. 14.21 − 7.8941   6.3159

14.2100
−7.8941
6.3159

Name ____________________

Unit I, Part C, Lesson 7, Quiz Form B
– Continued –

5. $.86 \div 43$ .02

$$\begin{array}{r} .02 \\ 43\overline{).86} \\ \underline{86} \\ 0 \end{array}$$

6. $13.648 + .293$ 13.941

$$\begin{array}{r} 13.648 \\ \underline{.293} \\ 13.941 \end{array}$$

7. $7.54 \cdot 1.328$ 10.01312

$$\begin{array}{r} 1.328 \\ \underline{7.54} \\ 5312 \\ 6640 \\ \underline{9296} \\ 10.01312 \end{array}$$

1.328 → 3 places to the right
7.54 → 2 places to the right
10.01312 → total of 5 places to the right

8. $15 - 2.64$ 12.36

$$\begin{array}{r} 15.00 \\ \underline{-2.64} \\ 12.36 \end{array}$$

Quiz Form A Name ______________________

Class ________ Date __________ Score ______

Unit I – The Structure of Mathematics
Part C – Further Investigation of Operation Symbols
Lesson 8 – Operations with Signed Numbers – Vectors and Absolute Value
Lesson 9 – Operations with Signed Numbers – Addition
Lesson 10 – Operations with Signed Numbers – Subtraction

Tell what signed number is represented by each of the following vectors.

1. 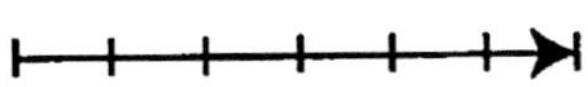 $+6$

This vector is 6 units long and points to the right, so it represents the signed number $^{+}6$.

2. 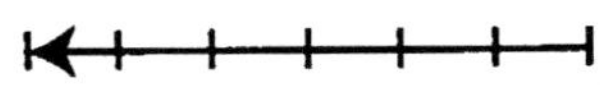 -6

This vector is 6 units long and points to the left, so it represents the signed number $^{-}6$.

Find the absolute value for each of the following as indicated.

3. $\left|{}^{+}17\right|$ 17

4. $\left|{}^{-}7\right|$ 7

5. $|5.3|$ 5.3

Find the answer for the following addition problem, using the given number line and vector addition.

6. $^{+}5 + {}^{+}2 + {}^{-}4$

$+3$

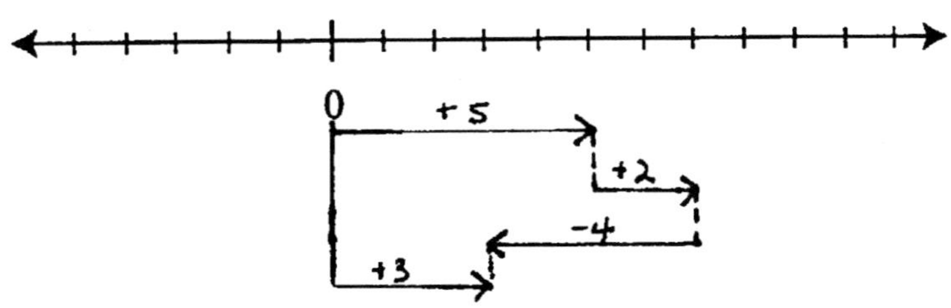

Find the answer mentally for each of the following problems.

7. $^{+}8 + {}^{+}3$ $+11$

8. $^{+}9 + {}^{-}4$ $+5$

9. $^{-}3 + {}^{-}6$ -9

Find the answer for each of the following problems by first <u>rewriting</u> the problem as a sum of signed numbers.

10. $9 - 7$

$9 + {}^{-}7$

2

11. $24 - 36$

$24 + {}^{-}36$

-12

12. $^{-}7 - {}^{+}8$

$-7 + {}^{-}8$

-15

Quiz Form B

Name ____________________

Class ________ Date __________ Score ______

Unit I – The Structure of Mathematics

Part C – Further Investigation of Operation Symbols

Lesson 8 – Operations with Signed Numbers – Vectors and Absolute Value

Lesson 9 – Operations with Signed Numbers – Addition

Lesson 10 – Operations with Signed Numbers – Subtraction

Tell what signed number is represented by each of the following vectors.

1. 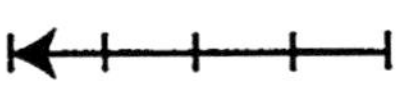 -4

This vector is 4 units long and points to the left, so it represents the signed number -4.

2. 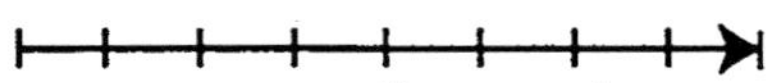 $+8$

This vector is 8 units long and points to the right, so it represents the signed number $+8$.

Find the absolute value for each of the following as indicated.

3. $\left|{}^{-}13\right|$

13

4. $\left|6.4\right|$

6.4

5. $\left|{}^{+}1\right|$

1

Find the answer for the following addition problem, using the given number line and vector addition.

6. ${}^{+}6 + {}^{-}4 + {}^{+}3$

$+5$

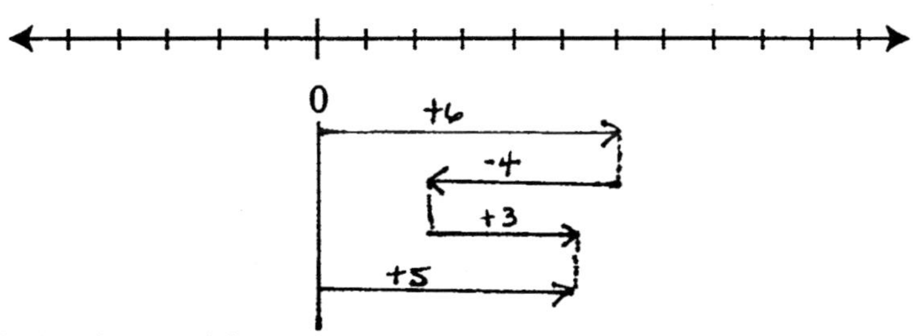

Find the answer mentally for each of the following problems.

7. ${}^{-}5 + {}^{-}7$

-12

8. ${}^{+}9 + {}^{+}8$

$+17$

9. ${}^{+}11 + {}^{-}4$

$+7$

Find the answer for each of the following problems by first rewriting the problem as a sum of signed numbers.

10. ${}^{-}6 - {}^{-}7$

$-6 + {}^{+}7$

$+1$

11. $12 - 9$

$12 + {}^{-}9$

3

12. ${}^{-}6 - {}^{+}10$

$-6 + {}^{-}10$

-16

Quiz Form A

Name ____________________

Class ________ Date ________ Score ______

Unit I – The Structure of Mathematics
Part C – Further Investigation of Operation Symbols

Lesson 11 – Operations with Signed Numbers – Multiplication and Division

Perform the indicated operation in each of the following. Be careful to attach the proper sign to your answer according to the guidelines of this lesson.

1. $^{+}6 \cdot {}^{+}21$ $\underline{\quad +126 \quad}$

2. $^{-}13 \cdot {}^{-}32$ $\underline{\quad +416 \quad}$

3. $^{+}64 \div {}^{-}32$ $\underline{\quad -2 \quad}$

4. $^{-}8 \cdot {}^{-}7$ $\underline{\quad +56 \quad}$

5. $^{-}32 \div {}^{+}4$ $\underline{\quad -8 \quad}$

6. $^{-}31 \cdot {}^{+}11$ $\underline{\quad -341 \quad}$

7. $^{+}7 \cdot {}^{-}15$ $\underline{\quad -105 \quad}$

8. $^{+}81 \div {}^{+}27$ $\underline{\quad +3 \quad}$

9. $^{-}72 \div {}^{-}9$ $\underline{\quad +8 \quad}$

10. $^{+}4 \div {}^{-}4$ $\underline{\quad -1 \quad}$

Quiz Form B

Name ____________________

Class ________ Date ________ Score ______

Unit I – The Structure of Mathematics
Part C – Further Investigation of Operation Symbols
Lesson 11 – Operations with Signed Numbers – Multiplication and Division

Perform the indicated operation in each of the following. Be careful to attach the proper sign to your answer according to the guidelines of this lesson.

1. $^{-}8 \div {}^{+}8$ ___ −1 ___

2. $^{-}64 \div {}^{-}16$ ___ +4 ___

3. $^{+}36 \div {}^{+}12$ ___ +3 ___

4. $^{+}8 \cdot {}^{-}17$ ___ −136 ___

5. $^{-}21 \cdot {}^{+}13$ ___ −273 ___

6. $^{-}44 \div {}^{+}11$ ___ −4 ___

7. $^{-}6 \cdot {}^{-}8$ ___ +48 ___

8. $^{+}81 \div {}^{-}27$ ___ −3 ___

9. $^{-}14 \cdot {}^{-}12$ ___ +168 ___

10. $^{+}7 \cdot {}^{+}31$ ___ +217 ___

Quiz Form A Name ______________________

Class ________ Date __________ Score ______

Unit I – The Structure of Mathematics
Part D – Further Investigation of Relation Symbols
Lesson 1 – Order of Numbers and the Number Line

For each of the following, label the appropriate points on the given number line by assigning the given numbers to those points as indicated.

1. $^{+}5$ assigned to point C

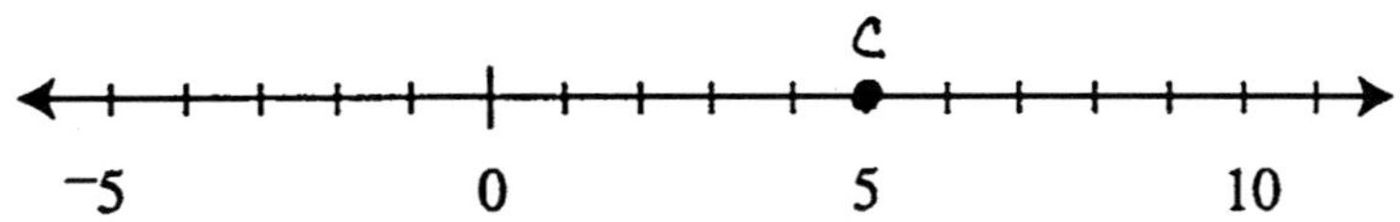

2. $^{-}4$ assigned to point T

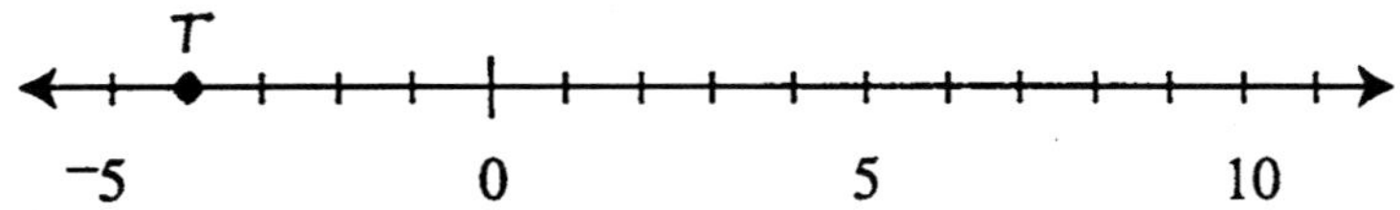

3. $^{-}3\frac{3}{4}$ assigned to point U

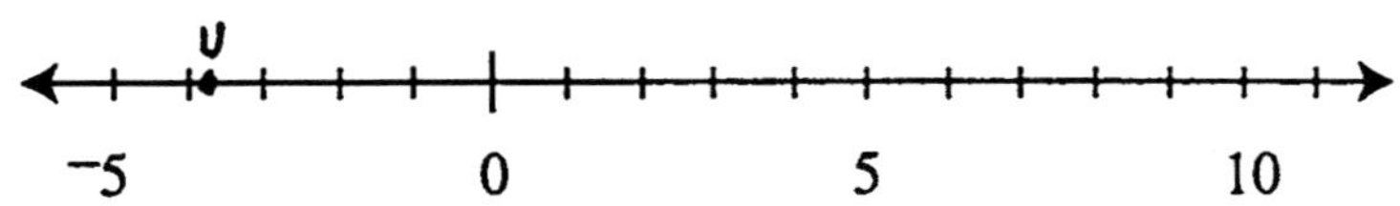

4. $^{+}1.75$ assigned to point X

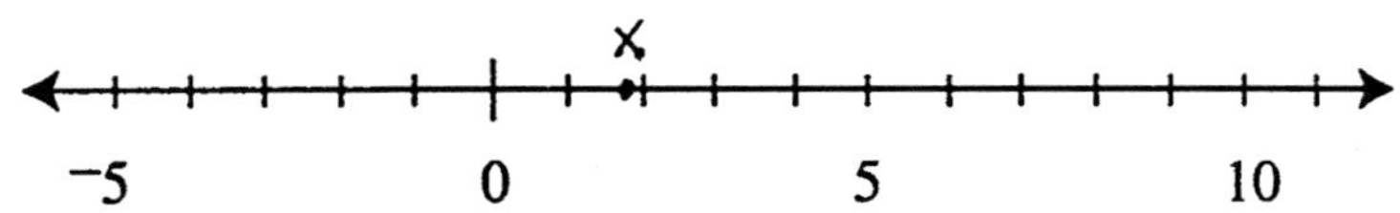

5. $^{-}2.2$ assigned to point A

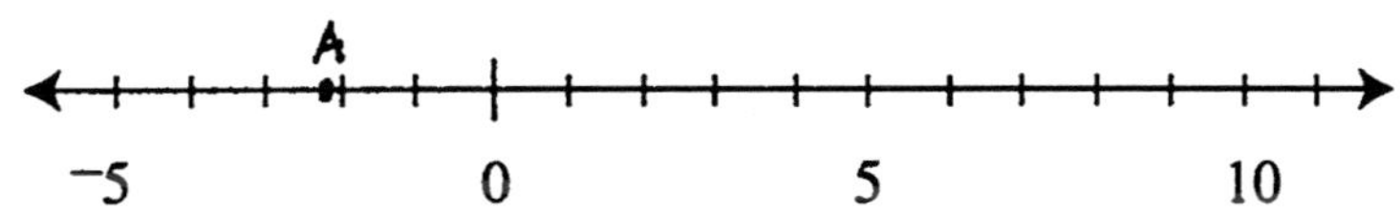

Use one of the three basic relation symbols (=, >, <) in each of the following to make a true statement of relationship.

6. $^{-}5$ __<__ 0 **7.** $^{-}9$ __<__ $^{-}5$ **8.** 0 __<__ $^{+}3$

9. $^{+}4$ __>__ $^{-}4$ **10.** $\frac{^{-}3}{7}$ __<__ $\frac{2}{9}$ **11.** $^{+}9$ __>__ $^{+}5$

12. $.05$ __<__ $.5$ **13.** $\frac{1}{3}$ __>__ $\frac{1}{4}$ **14.** $^{-}7.2$ __<__ $^{-}1.5$

15. If $^{+}5$ dollars means 5 dollars deposited in (or put into) a bank account, what does $^{-}5$ dollars mean?

$5.00 withdrawal

Quiz Form B

Name ______________________

Class ________ Date __________ Score ______

Unit I – The Structure of Mathematics
Part D – Further Investigation of Relation Symbols
Lesson 1 – Order of Numbers and the Number Line

For each of the following, label the appropriate points on the given number line by assigning the given numbers to those points as indicated.

1. $^{-}7$ assigned to point C

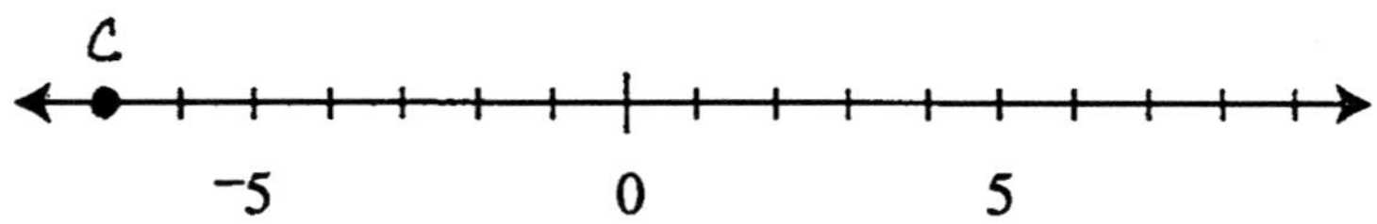

2. $^{+}3$ assigned to point T

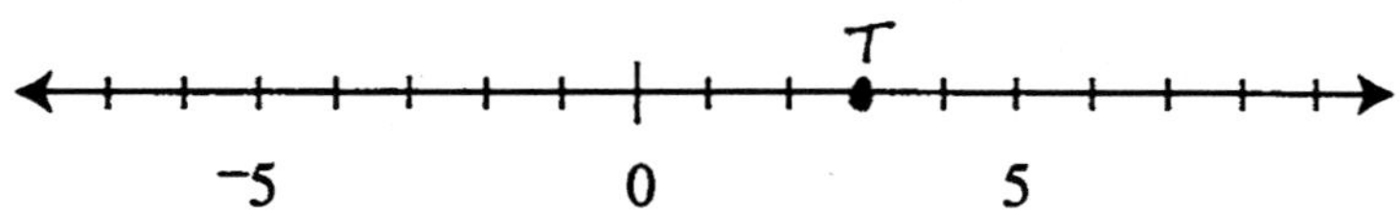

3. $4\frac{1}{4}$ assigned to point U

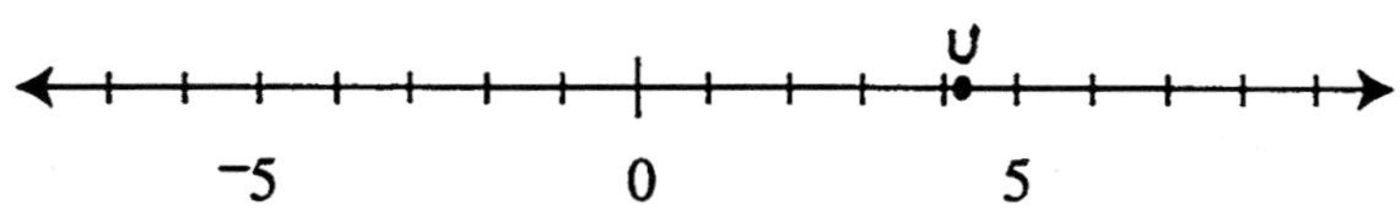

4. $^{-}1.25$ assigned to point X

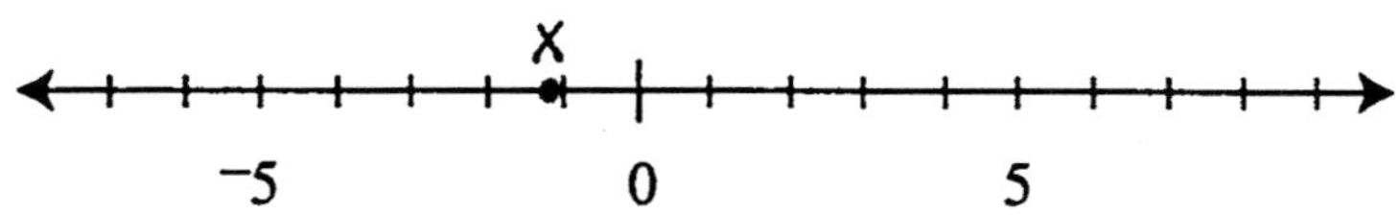

5. $^{-}3.4$ assigned to point A

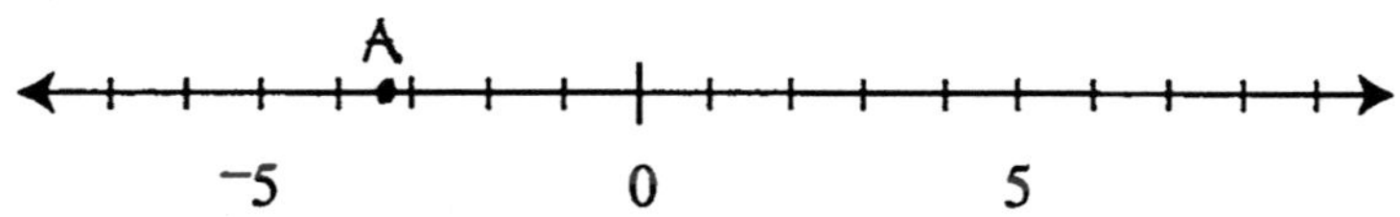

Use one of the three basic relation symbols (=, >, <) in each of the following to make a true statement of relationship.

6. $^{+}4$ __>__ 0

7. $^{-}7$ __>__ $^{-}11$

8. $^{+}2$ __<__ $^{+}6$

9. $\frac{2}{11}$ __>__ $\frac{^{-}3}{7}$

10. $\frac{1}{6}$ __<__ $\frac{1}{5}$

11. .7 __>__ .07

12. 0 __>__ $^{-}7$

13. $^{-}6.1$ __<__ $^{-}1.3$

14. 8 __>__ $^{-}8$

15. If $^{-}2$ means 2 miles below sea level, what does $^{+}2$ miles mean?

2 miles above sea level

Quiz Form A

Name ____________________

Class ________ Date __________ Score ______

Unit I – The Structure of Mathematics
Part D – Further Investigation of Relation Symbols
Lesson 2 – Properties of Equality
Lesson 3 – Properties of Inequality

State which property of equality or inequality justifies each of the following statements.

1. If $12 = 3x$, then $3x = 12$ Symmetric Property of Equality

2. If $y + 4 = 9$, and $9 = 5 + 4$, then $y + 4 = 5 + 4$ Transitive Property of Equality

3. $8 - 7xy = 8 - 7xy$ Reflexive Property of Equality

4. If $2 > {}^{-}1$, then $2 + 3 > {}^{-}1 + 3$ Addition Property of Inequality

5. If $x = y$, then $x - a = y - a$ Subtraction Property of Equality

6. If $3a = 9b$, then $a = 3b$ Multiplication Property of Equality

7. ${}^{-}6x > 12$ is the same as $x < {}^{-}2$ Multiplication Property of Inequality

In each of the following, perform the indicated operation on the given relation, indicating what the resulting sentence is.

8. $24 - 5 = 17 + 2$

Add ${}^{-}2$

$24 - 5 + {}^{-}2 = 17 + 2 + {}^{-}2$

$19 + {}^{-}2 = 19 + {}^{-}2$

$17 = 17$

9. $10 > 2$

Mult ${}^{-}\frac{1}{2}$

$-\frac{1}{2} \cdot 10 < -\frac{1}{2} \cdot 2$

${}^{-}5 < {}^{-}1$

Name ____________________

Unit I, Part D, Lessons 2 and 3, Quiz Form A
– Continued –

Extending your mind...

A relation R on a set A is an <u>equivalence</u> relation if it has the following properties.

Reflexive Property	For any element *a* of set A, *a* R *a*.
Symmetric Property	For all elements *a* and *b* of set A, if *a* R *b*, then *b* R *a*.
Transitive Property	For all elements *a*, *b*, and c of set A, if *a* R *b* and *b* R *c*, then *a* R *c*.

Equality on the set of all real numbers is reflexive, symmetric, and transitive. Therefore, it is an equivalence relation.

In each of the following, a relation and a set are given. Write yes if the relation is an equivalence relation on the given set. If it is not, write no, and tell which of the properties fail to hold true.

10. $>$ {all numbers} No

The Reflexive Property fails. (Examples will vary.)
For example, $3 > 3$ is not true.

The Symmetric Property fails. (Examples will vary.)
For example, if $5 > 3$, then $3 > 5$ is not true.

11. "is the brother of" {all boys in your school} No

The Reflexive Property fails. (Examples will vary.)
For example, John Robertson is the brother of John Robertson is not true.

Quiz Form B Name ______________________

Class ________ Date __________ Score ______

Unit I – The Structure of Mathematics
Part D – Further Investigation of Relation Symbols
Lesson 2 – Properties of Equality
Lesson 3 – Properties of Inequality

State which property of equality or inequality justifies each of the following statements.

1. If $5 < 7$, then $5 + 4 < 7 + 4$ Addition Property of Inequality

2. $17 + 3x = 17 + 3x$ Reflexive Property of Equality

3. If $^{-}3x < 12$, then $x > {}^{-}4$ Multiplication Property of Inequality

4. If $7 = 8 + {}^{-}1$, and $8 + {}^{-}1 = 21 \div 3$, then $7 = 21 \div 3$ Transitive Property of Equality

5. If $21 = 3x$, then $3x = 21$ Symmetric Property of Equality

6. If $4x = 20$, then $x = 5$ Multiplication Property of Equality

7. If $a = b$, then $a + x = b + x$ Addition Property of Equality

In each of the following, perform the indicated operations on the given relation, indicating what the resulting sentence is.

8. $14 > 7$

Mult $\frac{-1}{7}$ $-\frac{1}{7} \cdot 14 < -\frac{1}{7} \cdot 7$

$-2 < -1$

9. $19 + 7 = 25 + 1$

Subtract 1 $19 + 7 - 1 = 25 + 1 - 1$

$26 - 1 = 26 - 1$

$25 = 25$

Name ____________________

Unit I, Part D, Lessons 2 and 3, Quiz Form B
– Continued –

Extending your mind...

A relation R on a set A is an <u>equivalence</u> relation if it has the following properties.

Reflexive Property	For any element a of set A, a R a.
Symmetric Property	For all elements a and b of set A, if a R b, then b R a.
Transitive Property	For all elements a, b, and c of set A, if a R b and b R c, then a R c.

Equality on the set of all real numbers is reflexive, symmetric, and transitive. Therefore, it is an equivalence relation.

In each of the following, a relation and a set are given. Write yes if the relation is an equivalence relation on the given set. If it is not, write no, and tell which of the properties fail to hold true.

10. "$\neq$" {all numbers} No

The Reflexive Property fails. (Examples will vary.)
For example, $3 \neq 3$ is not true.

The Transitive Property fails. (Examples will vary.)
For example, if $2 \neq 3$ and $3 \neq 2$, then $2 \neq 2$ is not true.

11. "graduates the same year as" {any student in your school} Yes

All properties hold true!

Quiz Form A

Name ____________________

Class ________ Date ________ Score ______

Unit I – The Structure of Mathematics
Part E – Mathematical Models
Lesson 1 – The Mathematics of Sets

Given the following sets, find the union or intersection as indicated.

$A = \{1, 2, 3, 4, 5, 6, 7, \ldots 16, 17, 18\}$ $B = \{2, 4, 6, 8\}$

$C = \{3, 4, 5, 6\}$ $D = \{1, 2, 17, 18\}$

To find union we combine the members of both sets listing each member only once.

1. $C \cup D$ $\{1, 2, 3, 4, 5, 6, 17, 18\}$
2. $B \cup C$ $\{2, 3, 4, 5, 6, 8\}$

To find intersection we list all the members common to each set.

3. $D \cap B$ $\{2\}$
4. $C \cap B$ $\{4, 6\}$

Using the given sets for exercises 1-4, indicate if the following are true or false.

5. $C \subset A$ True, since C is contained in A, but does not contain all of the elements of A.
6. $(D \cap B) \subset C$ False, since $D \cap B = \{2\}$ and $\{2\}$ is not a proper subset of C.
7. $(B \cup C \cup D) \subset A$ True, since $B \cup C \cup D = \{1, 2, 3, 4, 5, 6, 8, 17, 18\}$ and this set is contained in A, but does not contain all the elements of A.
8. $D \cap C$ is the null set True, since there are no members common to D and C.
9. All of the given sets (A, B, C, and D) could be classified as finite sets True, since A, B, C, and D all have a definite number of members.

Name ____________________

Unit I, Part E, Lesson 1, Quiz Form A
– Continued –

10. If $A = \{a, 2\}$, $B = \{2, 4, c, d\}$ and $A \cap B = \{2\}$, draw a Venn Diagram that would represent the relationship between these sets.

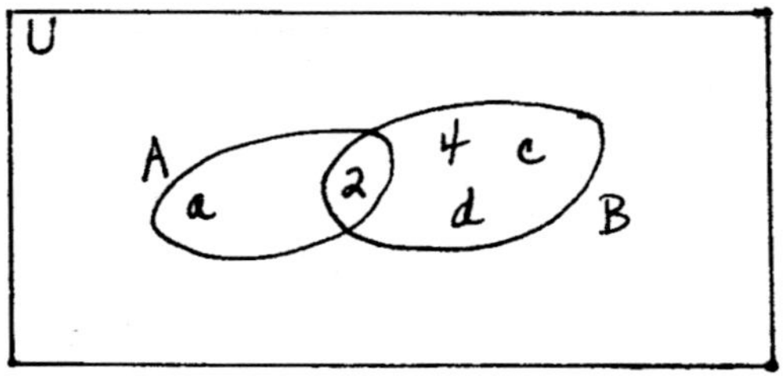

Quiz Form B

Name ______________________

Class ________ Date ________ Score ______

Unit I – The Structure of Mathematics
Part E – Mathematical Models

Lesson 1 – The Mathematics of Sets

Given the following sets, find the union or intersection as indicated.

$A = \{a, b, c, d, e, f, g, h, i, j, k, l, m, n, o, p, q\}$ $B = \{b, d, f, h\}$

$C = \{c, d, e, f\}$ $D = \{a, b, q\}$

1. $C \cup D$ — To find union we combine members of both sets listing each member only once. — $\{a, b, c, d, e, f, q\}$

2. $B \cup C$ — $\{b, c, d, e, f, h\}$

3. $D \cap B$ — To find intersection we list all the members common to each set. — $\{b\}$

4. $C \cap B$ — $\{d, f\}$

Using the given sets for exercises 1-4, indicate if the following are true or false.

5. $C \subset A$ — True, since C is contained in A, but does not contain all the members of A.

6. $(D \cap B) \subset C$ — False, since $D \cap B = \{b\}$ and $\{b\}$ is not a member of C.

7. A is finite — True, since A has a definite number of members.

8. $(B \cup C) \cap D$ is the null set — False, since $B \cup C = \{b, c, d, e, f, h\}$, and that set and D both have the member b in common.

9. $(C \cup D) \cap B$ contains one element — False, since $C \cup D = \{a, b, c, d, e, f, q\}$, and that set and B have the members b, d, and f in common.

Name ______________________________

Unit I, Part E, Lesson 1, Quiz Form B
– Continued –

10. $A = \{x, 5\}$, $B = \{5, 4, m, n\}$ and $A \cap B = \{5\}$, draw a Venn Diagram that would represent the relationship between these sets.

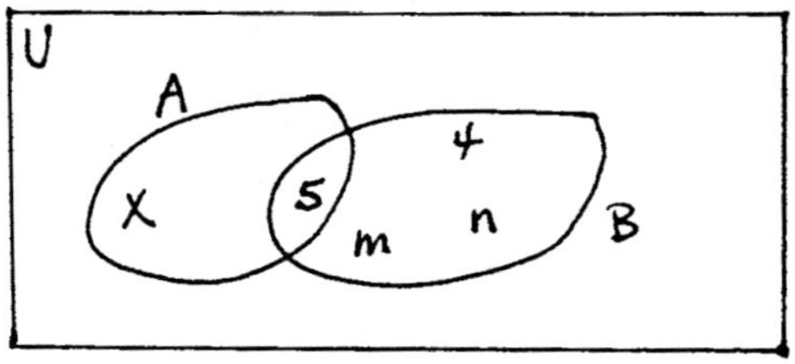

Quiz Form A **Name** ______________________

Class ________ **Date** __________ **Score** ______

Unit I – The Structure of Mathematics
Part E – Mathematical Models
Lesson 2 – The Mathematics of Functions

1. Given a domain $D = \{1, 3, 5, 7\}$ and a range $R = \{3, 7, 11, 15\}$, write an open phrase which defines a function.

   We can define a function from D to R by multiplying each member of D by 2 and then adding 1. The open phrase $2n + 1$ would define that function.

For each of the following functions, use the given domain, and find the range.

2. $f(\ ) = 3(\ )$ the domain is $D = \{^-1, ^-3, ^-5, ^-7\}$

   so the range is $R = \{^-3, ^-9, ^-15, ^-21\}$

   $f(^-1) = 3(^-1) = ^-3$
   $f(^-3) = 3(^-3) = ^-9$
   $f(^-5) = 3(^-5) = ^-15$
   $f(^-7) = 3(^-7) = ^-21$

3. $f(\ ) = \frac{(\ ) - 2}{3}$ the domain is $D = \{5, 8, 11, 14\}$

   so the range is $R = \{1, 2, 3, 4\}$

   $f(5) = \frac{5-2}{3} = \frac{3}{3} = 1$
   $f(8) = \frac{8-2}{3} = \frac{6}{3} = 2$
   $f(11) = \frac{11-2}{3} = \frac{9}{3} = 3$
   $f(14) = \frac{14-2}{3} = \frac{12}{3} = 4$

4. True or false?

   Using the definitions in this lesson, it could be said that all functions are relations, but all relations are not functions. Explain your answer.

   True. A relation is a set of ordered pairs; a function is a set of ordered pairs, no two of which have the same first component.

Quiz Form B Name ____________________

Class ________ Date ________ Score ______

Unit I – The Structure of Mathematics
Part E – Mathematical Models
Lesson 2 – The Mathematics of Functions

1. Given a domain $D = \{2, 4, 6, 8\}$ and a range $R = \{7, 11, 15, 19\}$, write an open phrase which defines a function.

   We can define a function from D to R by multiplying each member of D by 2, and then adding 3. The open phrase $2n+3$ would define that function.

For each of the following functions, use the given domain, and find the range.

2. $f(\ ) = {}^-2(\ )$ the domain is $D = \{{}^-1, {}^-3, {}^-5, {}^-7\}$

   $f({}^-1) = {}^-2({}^-1) = 2$

   $f({}^-3) = {}^-2({}^-3) = 6$ so the range is $R = \{2, 6, 10, 14\}$

   $f({}^-5) = {}^-2({}^-5) = 10$

   $f({}^-7) = {}^-2({}^-7) = 14$

3. $f(\ ) = \frac{(\ )+2}{4}$ the domain is $D = \{10, 14, 18, 22\}$

   $f(10) = \frac{10+2}{4} = \frac{12}{4} = 3$ so the range is $R = \{3, 4, 5, 6\}$

   $f(14) = \frac{14+2}{4} = \frac{16}{4} = 4$

   $f(18) = \frac{18+2}{4} = \frac{20}{4} = 5$

   $f(22) = \frac{22+2}{4} = \frac{24}{4} = 6$

4. True or false?

   Using the definitions in this lesson, if the domain was D = {all women in the world who are mothers}, and the range was R = {all women in the world}, then the relation defined by pairing each mother in the world with their daughter(s) would define a function. Explain your answer.

   False. It would be possible for a member of the domain to be paired with more than one member of the range.

Unit Test Form A

Name ______________________

Class __________ Date __________ Score ________

Unit I – The Structure of Mathematics

In problems 1-5, state the five parts of mathematical speech, and give an example of each.

(Examples will vary.)

(A-1)

1. Number Symbols — Example: 3
2. Operation Symbols — Example: +
3. Relation Symbols — Example: =
4. Grouping Symbols — Example: ( )
5. Placeholder Symbols — Example: n

For each expression in problems 6-10, tell (a) whether the expressions are open or closed, sentences or phrases, and (b) do whatever you can with each. If you need a domain, use $\{1, 2, 3\}$. If you need a replacement set, use $\{4, 5, 6\}$.

(A-2)

6. $m+11$
   (a). open phrase (placeholder, but no relation symbol)
   (b). $m+11$; $1+11=12$; $2+11=13$; $3+11=14$; $\{12, 13, 14\}$

7. $9 \cdot 7+0$
   (a). closed phrase (no placeholder and no relation symbol)
   (b). $9\cdot 7+0$; $63+0$; 63 — 63

8. $3=6-x$
   (a). open sentence (placeholder and relation symbol)
   (b). $3=6-x$; $3=6-4$ F; $3=6-5$ F; $3=6-6$ F — $\{\ \}$

9. 27,613
   (a). closed phrase (no placeholder and no relation symbol)
   (b). 27,613

10. $\dfrac{30-12}{3}=7-2$
    (a). closed sentence (relation symbol, but no placeholder)
    (b). $\frac{30-12}{3}=7-2$; $\frac{18}{3}=7-2$; $6=5$ False

(B-1) 11. Tell which of the following are rational numbers by circling the appropriate numbers.

$\sqrt{29}$ $\quad \pi \quad$ ($^-15$) can be written as $\frac{-15}{1}$. $\quad$ ($\frac{6}{7}$) $\quad$ ($5.\bar{2}$) can be written as $\frac{47}{9}$. $\quad$ ($\sqrt{36}$) can be written as $\frac{6}{1}$.

Name ______________________________

Unit I – The Structure of Mathematics, Unit Test Form A – Continued, Page 2 –

(B-3)
12. Express $\frac{3}{16}$ as a decimal. Show your work.

$$16\overline{)3.0000} = .1875$$

$\frac{3}{16} = .1875$

(B-2)
13. Complete the following to find the equivalent form.

$$\frac{3}{8} = \frac{27}{72} \qquad \frac{3}{8} = \frac{3}{8}\cdot\frac{9}{9} = \frac{27}{72}$$

(B-2)
14. Simplify the following by writing both the numerator and the denominator in prime factored form, and then finding all the "1's" you can.

$$\frac{42}{90} = \frac{2\cdot 3\cdot 7}{2\cdot 3\cdot 3\cdot 5} = \frac{7}{15}$$

(B-3)
15. Express $\frac{1}{11}$ as a decimal. Show your work.

$$11\overline{)1.0000} = .0909$$

This process will continue forever, so we write $.\overline{09}$.

(B-4)
16. Express .45 as a fraction. Show your work.

.45 is read as "45 hundredths" which is represented by $\frac{45}{100}$ or $\frac{9}{20}$.

(B-5)
17. Express the number 1.2 as a percent.

Move the decimal point 2 places to the right, and attach a percent sign.

1.20   So, 1.2 = 120%

(B-4)
18. Express $.\overline{39}$ as a fraction. Show your work.

$x = .\overline{39}$

$$100x = 39.\overline{39}$$
$$-1x = -.\overline{39}$$
$$99x = 39$$
$$x = \frac{39}{99} = \frac{3\cdot 13}{3\cdot 33} = \frac{13}{33}$$

For problems 19-21, factor each number as a product of primes. Then, in problem 22, find the greatest common factor of the three numbers.

(B-6)
19. 18 $18 = 2\cdot 3\cdot 3$

(B-6)
20. 30 $30 = 2\cdot 3\cdot 5$

(B-6)
21. 42 $42 = 2\cdot 3\cdot 7$

(B-8)
22. The greatest common factor is: GCF = $2\cdot 3 = 6$

Name ______________________________

Unit I – The Structure of Mathematics, Unit Test Form A
– Continued, Page 3 –

Simplify each of the expressions in problems 23 and 24.

(C-1)
23. $(^{-}12)(5)-6$

$^{-}60 - 6$

$^{-}60 + {^{-}6}$

$^{-}66$

(C-1)
24. $62\cdot 8-(5-64\div 4)$

$62\cdot 8-(5-16)$

$62\cdot 8-(^{-}11)$

$496-(^{-}11)$

$496+11$

507

For problems 25-27, find $-a$ when a is given. (In other words, find the opposite, or additive inverse, of each number.)

(C-3)
25. $a={^{-}21}$

$-a=21$

(C-3)
26. $a=8$

$-a={^{-}8}$

(C-3)
27. $a=\frac{5}{b}$

$-a=-\frac{5}{b}$

Rewrite each expression in problems 28 and 29 using the indicated property of operations.

(C-2)
28. $m\cdot p$ Commutative Law

$p\cdot m$

(C-2)
29. $7+(a+b)$ Associative Law

$(7+a)+b$

Find each sum in problems 30-32.

(C-9)
30. $17+{^{-}24}$

-7

(C-7, 9)
31. ${^{-}3.6}+{^{-}7.8}$

-11.4

(C-5, 9)
32. $\frac{^{-}6}{7}+\frac{1}{2}$

The LCM of 7 and 2 is 7·2 or 14.

$-\frac{6}{7}\cdot 1+\frac{1}{2}\cdot 1$

$-\frac{6}{7}\cdot\frac{2}{2}+\frac{1}{2}\cdot\frac{7}{7}$

$\frac{-12}{14}+\frac{7}{14}$

$\frac{^{-}12+7}{14}$

$\frac{^{-}5}{14}$

Name ______________________________

Unit I – The Structure of Mathematics, Unit Test Form A
– Continued, Page 4 –

Find each product in problems 33-35.

(c-11)

33. $^{-}3(2)(^{-}1)(^{-}6)$

$-6(-1)(-6)$

$6(-6)$

-36

(c-7,11)

34. $^{-}15(^{-}4.5)$

$^{-}15 \rightarrow$ no places to the right

$-4.5 \rightarrow$ 1 place to the right

75

60

$+67.5 \rightarrow$ total of 1 place to the right

(c-4,11)

35. $\frac{7}{8}\left(\frac{^{-}3}{5}\right) = \frac{7 \cdot {}^{-}3}{8 \cdot 5} = \frac{^{-}21}{40}$

Find each difference in problems 36-38.

(c-7,10)

36. $^{-}17.1 - 6.3$

$^{-}17.1 + {}^{-}6.3$

$^{-}23.4$

(c-10)

37. $^{-}9 - (^{-}3)$

$^{-}9 + {}^{+}3$

$^{-}6$

(c-5,10)

38. $\frac{7}{4} - \left(\frac{^{-}3}{2}\right)$

$\frac{7}{4} + \frac{^{+}3}{2}$

LCM of 4 and 2 is 4.

$\frac{7}{4} + \frac{3}{2} \cdot 1$

$\frac{7}{4} + \frac{3}{2} \cdot \frac{2}{2}$

$\frac{7}{4} + \frac{6}{4}$

$\frac{7+6}{4} \rightarrow \frac{13}{4}$

Find each quotient in problems 39-41.

(c-7,11)

39. $35.7 \div (^{-}8.5)$

$-8.5\overline{)35.70}$ = -4.2

340

170

170

0

(c-6,11)

40. $\frac{11}{12} \div \left(\frac{^{-}4}{5}\right)$

$\frac{11}{12} \cdot \frac{^{-}5}{4}$

$\frac{11 \cdot {}^{-}5}{12 \cdot 4}$

$\frac{^{-}55}{48}$

(c-7,11)

41. $\frac{^{-}6.6}{11}$

$11\overline{)-6.6}$ = $-.6$

66

0

Name ______________________________

Unit I – The Structure of Mathematics, Unit Test Form A
– Continued, Page 5 –

Simplify each of the expressions in problems 42-44.

(C-8)
42. $|48|$

48

(C-8)
43. $|{}^-4.18|$

4.18

(C-8)
44. $|18-18|$

$|0|$

0

Evaluate and simplify each expression in problems 45 and 46.

(A-2)
45. $6a-2b$ for $a=3$ and $b=1$.

$6(3)-2(1)$

$18-2$

16

(A-2)
46. $\dfrac{x-y}{3}$ for $x=17$ and $y=5$.

$\dfrac{17-5}{3}$

$\dfrac{12}{3}$

4

For problems 47 and 48, write a true statement, using " < " or " > " to replace □.

(D-1)
47. $^-8$ [<] 6

(D-1)
48. $^-5$ [<] $^-2$

Use the accompanying number lines in problems 49 and 50 to show each of the following.

(D-1)
49. $x={}^-2$

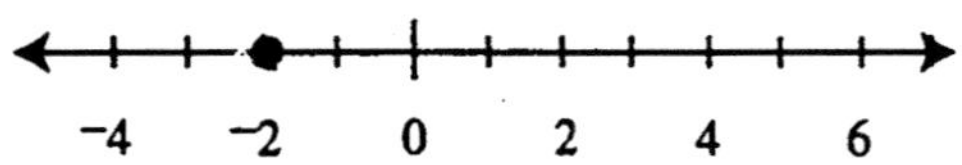

(C-9)
50. Add $5+{}^-2$ using vectors.

$5+{}^-2={}^+3$

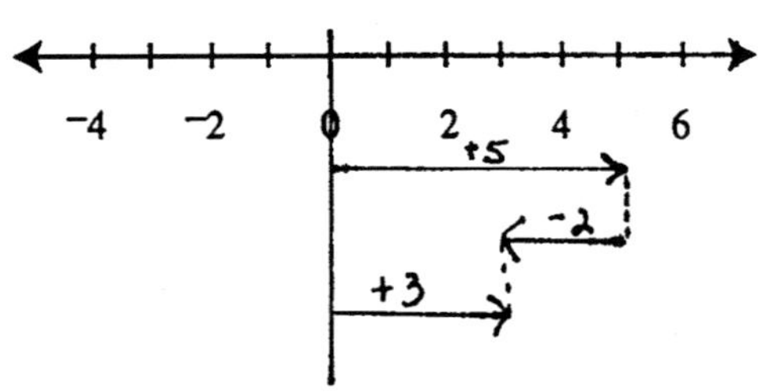

Name ______________________

Unit I – The Structure of Mathematics, Unit Test Form A
– Continued, Page 6 –

For each of the true statements in problems 51-54, state which property of equality or inequality is illustrated.

(D-2)
51. If $9 = 6 + 3$, then $6 + 3 = 9$
Symmetric Property of Equality

(D-2)
52. $^-7 = {}^-7$
Reflexive Property of Equality

(D-3)
53. If $x > y$, then $x + a > y + a$
Addition Property of Inequality

(D-3)
54. If $^-12 < 4$, then $3 > {}^-1$
Multiplication Property of Inequality

$^-12 < 4$
$-\frac{1}{4} \cdot {}^-12 > -\frac{1}{4} \cdot 4$
$3 > {}^-1$

(E-1)
55. Find the union of $\{1, 2, 3, 4\}$ and $\{3, 4, 5, 6\}$.
To find union we combine the members of both sets listing each member only once.
$\{1, 2, 3, 4, 5, 6\}$

(E-1)
56. Find the intersection of $\{1, 2, 3, 4\}$ and $\{3, 4, 5, 6\}$.
To find intersection we list all the members common to each set.
$\{3, 4\}$

(E-2)
57. Given a function defined as $f(\ \ ) = {}^-7(\ \ ) + 1$, and a domain $D = \{3, 5, 7, 9\}$, write an explanation of the process you would use to find the range. Then, actually find the range.

Answers will vary.
Each element of the domain must be substituted in the function to find each element of the range.

$f(\ ) = {}^-7(\ ) + 1$

If we substitute 3, $f(3) = {}^-7(3) + 1$ or $^-20$.
If we substitute 5, $f(5) = {}^-7(5) + 1$ or $^-34$.
If we substitute 7, $f(7) = {}^-7(7) + 1$ or $^-48$.
If we substitute 9, $f(9) = {}^-7(9) + 1$ or $^-62$.

The range is $\{^-20, {}^-34, {}^-48, {}^-62\}$.

Unit Test Form B

Name ______________________

Class __________ Date __________ Score ______

Unit I – The Structure of Mathematics

In problems 1-4, state the four types of mathematical expressions, and give an example of each.

Examples will vary.

(A-2)

1. open phrase — Example: $n+7$
2. closed phrase — Example: $7\cdot 8+4$
3. open sentence — Example: $2x-1=8$
4. closed sentence — Example: $5\cdot 4+6=36\div 12$

In problems 5-10, translate each of the English phrases into mathematical expressions.

(A-3)

5. 85% of x — $.85x$
6. three times d — $3\cdot d$ or $3d$ or $3\times d$
7. 19 more than c — $19+c$
8. 11 less than m — $m-11$
9. twice y — $2\cdot y$ or $2y$ or $2\times y$
10. the square of a, divided by 8 — $a^2\div 8$ or $\frac{a^2}{8}$

(B-1)

11. Tell which of the following are irrational numbers by circling the appropriate numbers.

$\sqrt{121}$ — rational; can be written as $\frac{11}{1}$

(π) — circled

$\frac{^-4}{3}$ — rational

$9.\overline{7}$ — rational; can be written as $\frac{88}{9}$.

$^-15$ — rational; can be written as $\frac{^-15}{1}$.

($\sqrt{6}$) — circled

Name ______________________________

Unit I – The Structure of Mathematics, Unit Test Form B – Continued, Page 2 –

(B-2)

12. Complete the following to find the equivalent form.

$$\frac{11}{3}=\frac{165}{45} \qquad \frac{11}{3}\cdot\frac{15}{15}=\frac{165}{45}$$

(B-2)

13. Simplify the following by writing both the numerator and the denominator in prime factored form, and then finding all the "1's" you can.

$$\frac{30}{105}=\frac{2\cdot\cancel{3}\cdot\cancel{5}}{\cancel{3}\cdot\cancel{5}\cdot7}=\frac{2}{7}$$

(B-3)

14. Express $\frac{3}{8}$ as a decimal. Show your work.

.375

$$8\overline{)3.000} = .375$$

24, 60, 56, 40, 40, 0

(B-3)

15. Express $\frac{3}{11}$ as a decimal. Show your work.

$.\overline{27}$

$$11\overline{)3.0000} = .2727\ldots$$

22, 80, 77, 30, 22, 80, 77, 3

This process will continue forever, so we write $.\overline{27}$.

(B-4)

16. Express .52 as a fraction. Show your work.

.52 is read as "fifty-two hundredths" which is represented by $\frac{52}{100}$ or $\frac{13}{25}$.

(B-4)

17. Express $.\overline{21}$ as a fraction. Show your work.

$\frac{21}{99}$ or $\frac{7}{33}$

$$100f = 21.\overline{21}$$
$$-1f = -.\overline{21}$$
$$99f = 21$$
$$f = \frac{21}{99} = \frac{\cancel{3}\cdot7}{\cancel{3}\cdot33} = \frac{7}{33}$$

(B-5)

18. Express the number .14 as a percent.

Move the decimal point two places to the right, and attach a percent sign.

.14 + % So, .14 = 14%

For problems 19-21, factor each number as a product of primes. Then, in problem 22, find the least common multiple of the three numbers.

(B-6)

19. 9 = $3\cdot3$

(B-6)

20. 6 = $2\cdot3$

(B-6)

21. 15 = $3\cdot5$

(B-7)

22. The least common multiple is: $2\cdot3\cdot3\cdot5$ or 90

Name ______________________________

Unit I – The Structure of Mathematics, Unit Test Form B
– Continued, Page 3 –

Simplify each of the expressions in problems 23 and 24.

(C-1)
23. $53-4\cdot 28+34$

$53-112+34$

$^-59+34$

$^-25$

(C-1)
24. $(6)(^-13)\div 3$

$^-78\div 3$

$^-26$

For problems 25-27, find $-a$ when a is given. (In other words, find the opposite, or additive inverse, of each number.)

(C-3)
25. $a=5$

$-a={}^-5$

(C-3)
26. $a={}^-24$

$-a=24$

(C-3)
27. $a=6.8$

$-a={}^-6.8$

Rewrite each expression in problems 28 and 29 using the indicated property of operations.

(C-2)
28. $d+(e+f)$ Associative Law

$(d+e)+f$

(C-2)
29. $a+b$ Commutative Law

$b+a$

Find each sum in problems 30-32.

(C-5, 9) The LCM of 7 and 3 is 7·3 or 21.

30. $\frac{^-5}{7}+\frac{1}{3}$

$\frac{^-5}{7}\cdot 1+\frac{1}{3}\cdot 1$

$\frac{^-5}{7}\cdot\frac{3}{3}+\frac{1}{3}\cdot\frac{7}{7}$

$\frac{^-15}{21}+\frac{7}{21}$

$\frac{^-15+7}{21}$

$\frac{^-8}{21}$

(C-7, 9)
31. $^-4.7+(^-5.2)$

$^-4.7$
$+\ ^-5.2$
$^-9.9$

(C-9)
32. $15+{}^-23$

$^-8$

Name ______________________________

Unit I – The Structure of Mathematics, Unit Test Form B – Continued, Page 4 –

Find each product in problems 33-35.

(C-7,11)
33. $\left(^{-}4.1\right)\left(^{-}1.2\right)$

$^{-}4.1 \rightarrow$ 1 place to the right
$^{-}1.2 \rightarrow$ 1 place to the right
82
41
$+4.92 \rightarrow$ total of 2 places to the right

(C-4,11)
34. $^{-}\frac{2}{3}\left(\frac{6}{7}\right) = \frac{^{-}2\cdot 6}{3\cdot 7} = \frac{^{-}2\cdot 2\cdot 3}{3\cdot 7} = \frac{^{-}4}{7}$

(C-11)
35. $^{-}6(4)\left(^{-}3\right)(5)$
$^{-}24\,(^{-}3)(5)$
$72(5)$
360

Find each difference in problems 36-38.

(C-7,10)
36. $^{-}40.1 - 6.8$
$^{-}40.1 + {}^{-}6.8$
$^{-}46.9$

(C-10)
37. $^{-}17 - \left(^{-}5\right)$
$^{-}17 + 5$
$^{-}12$

(C-5,10)
38. $\frac{8}{9} - \left(\frac{^{-}2}{3}\right)$
$\frac{8}{9} + \frac{2}{3}$
LCM of 9 and 3 is 9.

$\frac{8}{9} + \frac{2}{3}\cdot 1$
$\frac{8}{9} + \frac{2}{3}\cdot\frac{3}{3}$
$\frac{8}{9} + \frac{6}{9}$
$\frac{8+6}{9}$ or $\frac{14}{9}$ or $1\frac{5}{9}$

Find each quotient in problems 39-41.

(C-6,11)
39. $\left(\frac{^{-}7}{3}\right) \div \frac{5}{6}$
$\frac{^{-}7}{3}\cdot\frac{6}{5}$
$\frac{^{-}7\cdot 6}{3\cdot 5}$
$\frac{^{-}7\cdot 2\cdot 3}{3\cdot 5} = {}^{-}\frac{14}{5}$ or $^{-}2\frac{4}{5}$

(C-7,11)
40. $8.64 \div \left(^{-}3.6\right)$
$^{-}2.4$
$^{-}3.6\overline{)\,8.64}$
72
144
144
0

(C-7,11)
41. $\frac{^{-}55.8}{^{-}6.2}$
$^{+}9.$
$^{-}6.2\overline{)\,^{-}55.8}$
558
0

Name ______________________________

Unit I – The Structure of Mathematics, Unit Test Form B – Continued, Page 5 –

Simplify each of the expressions in problems 42-44.

(C-8)
42. $\left| \frac{^-3}{11} \right|$

$\frac{3}{11}$

(C-8)
43. $|26-26|$

$|0|$

0

(C-8)
44. $|17.2|$

17.2

Evaluate and simplify each expression in problems 45 and 46.

(A-2)
45. $\frac{5a}{2b}$ for $a=2$ and $b=3$.

$\frac{5\cdot 2}{2\cdot 3} = \frac{5}{3}$ or $1\frac{2}{3}$

(A-2)
46. $\frac{4x-y}{3}$ for $x=5$ and $y=2$.

$\frac{4(5)-2}{3}$

$\frac{20-2}{3}$

$\frac{18}{3}$

6

For problems 47 and 48, write a true statement using " < " or " > " to replace ☐.

(D-1)
47. 5 [>] $^-4$

(D-1)
48. $^-7$ [>] $^-13$

Use the accompanying number lines in problems 49 and 50 to show each of the following.

(D-1)
49. $x=4$

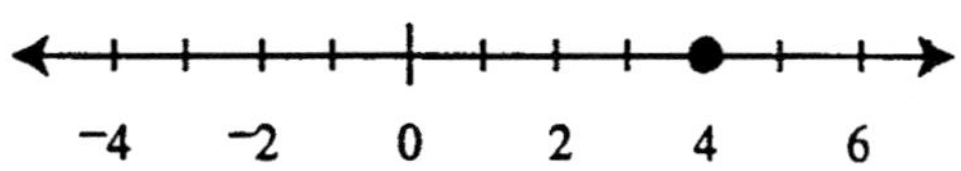

(C-9)
50. Add $^-3+5$ using vectors.

$-3+5=+2$

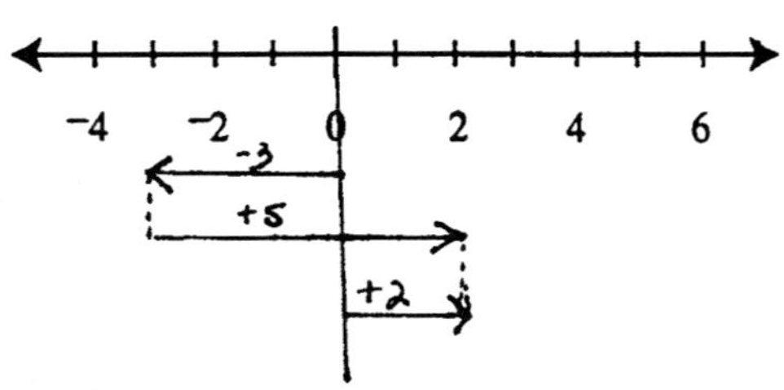

Name ______________________________

Unit I – The Structure of Mathematics, Unit Test Form B – Continued, Page 6 –

For each of the true statements in problems 51-54, state which property of equality or inequality is illustrated.

(D-2)
51. If $a = b$, then $a + c = b + c$

Addition Property of Equality

(D-2)
52. If $a = b$, then $b = a$

Symmetric Property of Equality

(D-3)
53. If $12 > 4$, then $3 > 1$

Multiplication Property of Inequality

$12 > 4$
$\frac{1}{4}(12) > \frac{1}{4}(4)$
$3 > 1$

(D-2)
54. If $5 = 4 + 1$, and $4 + 1 = 10 \div 2$, then $5 = 10 \div 2$

Transitive Property of Equality

(E-1)
55. Find the intersection of $\{a, b, c, d\}$ and $\{b, c, e, f\}$.

To find intersection we list all the members common to each set.

$\{b, c\}$

(E-1)
56. Find the union of $\{a, b, c, d\}$ and $\{b, c, e, f\}$.

To find union we combine the members of both sets listing each member only once.

$\{a, b, c, d, e, f\}$

(E-2)
57. Given a function defined as $f(\ ) = 5(\ ) - 2$, and a domain $D = \{3, 5, 7, 9\}$, write an explanation of the process you would use to find the range. Then actually find the range.

Answers will vary. For example, each element of the domain must be substituted in the function to find each element of the range.

$f(\ ) = 5(\ ) - 2$

If we substitute 3, $f(3) = 5(3) - 2$ or 13.
If we substitute 5, $f(5) = 5(5) - 2$ or 23.
If we substitute 7, $f(7) = 5(7) - 2$ or 33.
If we substitute 9, $f(9) = 5(9) - 2$ or 43.

The range is $\{13, 23, 33, 43\}$.

Index for Student WorkText

A

Abscissa 150
Absolute Value 44, 47, 53, 108, 110, 114, 172
Addition
  complex numbers 410
  of fractions 338
  property of equality 60, 194
  property of inequality 63
  radicals 387
  vector 47
Addition Property for Radicals 387
Additive Inverse Property 33
Algebraic Expression
  degree of 276
Algebraic Expressions 275, 278
And 99, 110
Angles
  complementary 129, 218, 254
  supplementary 129, 218, 254
Area 326
Associative Property 30
Asymptotes of a Hyperbola 496, 502
Axes of Symmetry of an Ellipse 489, 493
Axis
  coordinate 150
  horizontal 150
  of symmetry of ellipse 489, 493
  of symmetry of hyperbola 496, 502
  symmetry 462, 466, 470, 475
  vertical 150
Axis of Symmetry for a Hyperbola 496, 502

B

Base 263
Binomial 278
Branches of a Hyperbola 496, 502

C

Cartesian Coordinate System 403
Center of a Circle 482, 485
Center of a Hyperbola 496, 502
Center of an Ellipse 489, 493
Characteristic 535, 542
Circle 460, 482, 485
  center 482, 485
  radius 482, 485
Closed Phrase 2, 26
Closed Sentence 2
Coefficient 276
Combined Variation 371
Commission 137, 227
Common Factors of Several Numbers 24
Common Logarithms 535, 541
Common Multiples 22
Commutative Property 30
Compatible Radicals
  for addition and subtraction 387
  for division 385
  for multiplication 379, 381
Complementary Angles 129, 218, 254
Completing the Square 422
Complex Forms 342
Complex Fraction 39, 342
Complex Number 408
Complex Numbers
  adding 410
  dividing 415
  multiplying 412
  subtracting 410
Composite Number 20
Compound Mathematical Sentence 99, 104
Conjugate 392, 415
Conjunction 99, 104, 188, 191
Consecutive Even Integers 323
Consecutive Integers 123, 323
Consecutive Odd Integers 323
Consistent System 238, 244
Coordinate Axes 150
Coordinate of a Point 57
Critical Point 432
Cube Root 377
Cubes
  difference of 308
  sum of 308

D

Decimal 10, 12, 14
  changing to fraction 14
  repeating 12, 14
  terminating 12, 14
Decimal Numbers 41
Deduction 227
Deductions 137
Degree
  of a term 276
  of a variable 276
  of an algebraic expression 276
Denominator 10, 34, 37
  rationalize 392, 415
Dependent System 244
Difference of Cubes 308
Difference of Squares 295
Difference of Two Perfect Cubes 309
Difference of Two Perfect Squares 296
Difference of Two Squares 392, 415
Direct Variation 366
Discount 137, 226
Distance Formula 403
Distributive Property 30, 87, 89
Division
  complex numbers 415
  of fractions 336
  radicals 385
  synthetic 315

E

Ellipse 460, 488, 493
  axes of symmetry 489, 493
  center 489, 493
Equality 66
Equation
  radical 395, 398
Equations
  algebraic solution for 199
  exponential 546, 550
  graphing 153, 156, 188
  linear 141, 188, 194, 199, 202, 206
  literal 317, 353
  logarithm 550
  quadratic 419
  slope-intercept form 177, 179, 182
  solution sets 141
Equations and Inequalities 73, 76, 79, 83
  complications on 87, 89, 91, 94
  solving by factoring 288, 292, 295, 298, 301, 305, 308, 312, 315, 317
  special cases of 96, 97
  systems of 99, 104, 108, 110, 114
Equivalent Forms 10
Eratosthenes' Sieve 20
Evaluate a Closed Phrase 26
Exponent 263
  equation 546, 550
  function 521, 524, 527, 530
  rational 375
Exponent Notation 263, 265, 268, 271, 273
Exponents
  fractions 375
Expression
  radical 375
Extraneous Roots 395, 398

F

Factoring 20
  general strategy 312
Factors of a Number 24
Finite Set 66
Fraction 10, 34, 37
Fractions
  adding 37, 338
  as exponents 375
  complex 39, 342
  dividing 39, 336
  multiplying 34, 334
  subtracting 37, 338
Function 70
Functions
  exponential 521, 524, 527, 530
  inverse 524, 527, 530
  logarithmic 527, 530

G

General Factoring Strategy 312
General Trinomial 301
Geometric Figures 129, 218, 254, 326
Graph of a Number 57
Graphing
  intercepts 159
  of absolute value 172
  of equations 153, 156
  of first degree relations 164, 168
  terminology 149
Greatest Common Factor 24, 292
Grouping Symbols 1, 87

H

Horizontal Axis 150
Hyperbola 461, 495, 502
  asymptotes 496, 502
  axis of symmetry 496, 502
  branches 496, 502
  center 496, 502
  vertices 496, 502

I

Identity Element 32
  for addition 32
  for division 33
  for multiplication 32
  for subtraction 32
Identity Property 32
  of addition 32
  of division 32
  of multiplication 32
  of subtraction 32
Imaginary Number 408
Inconsistent System 242
Independent System 238, 242
Index 375
Inequalities
  linear 147, 191
  solution sets 147
Infinite Set 66
Integers 7, 57
  consecutive 123, 323
  consecutive even 123, 323
  consecutive odd 123, 323
Intercept 156, 159, 164, 168, 172
Intercepts 478
  graphing 159
Interest 137, 227
Intersection 66
Inverse
  element for multiplication 33
  elements for addition 33
  functions 524, 527, 530
Inverse Variation 368
Irrational Numbers 8
Is a Proper Subset of 67
Is a Subset of 66
Isolate a Radical 395, 398

J

Joint Variation 370

L

Least Common Multiple 22, 37, 338
Like Terms 89
Linear Equation 141
Linear Equations
  system of simultaneous 188, 194, 199, 202, 206
Linear Inequalities
  system of simultaneous 191
Linear Inequality 147
Lines
  parallel 185
  perpendicular 185
Literal Equation 317, 353
Logarithm 527, 530
  function 527, 530
Logarithms
  common 535, 541
  equation 550
  power property 532
  product property 532
  property of equality 546
  quotient property 532
  table of common 557, 558

M

Mantissa 535, 542
Markup 137, 226
Mathematical Models 66, 70
Members of a Set 66
Mixed Variation 370
Monomial 278
Multiples of a Number 22
Multiplication
  complex numbers 412
  of fractions 334
  of radicals 379
  of reciprocals 79
  property of equality 60, 194
  property of inequality 63
  property of zero 33
Multiplicative Inverse Property 33
My Dear Aunt Sally 26

N

Natural Numbers 7, 57
Null Set 66
Number Line 44, 47, 57
Numbers
  common factors of 24
  common multiples of 22
  complex 408
  composite 20
  decimal 41
  factoring with primes 20
  factors of 24
  graph of 57
  greatest common factors of 24
  identifying types 7
  imaginary 408
  irrational 8
  least common multiple of 22
  multiples of 22
  natural 7, 57
  prime 20
  rational 8, 57
  real 8
  relatively prime 24
  symbols 1
  whole 7, 57
Numerator 10, 34, 37

O

Open Phrase 2
Open Sentence 2
  solution set for 149, 153, 156, 159
Operation
  closure 7, 29
  existence 7, 29
  order of 26
  properties 29
  symbols 1, 26, 29, 32, 34, 37, 39, 41, 44, 47, 51, 53
  uniqueness 7, 29
  well-defined 7, 29
  with decimals 41
  with one 32

with signed numbers
addition 47
multiplication and division 53
subtraction 51
vectors and absolute value 44
with zero 32
Operation Symbols 1
Opposite Of 76, 83, 91
Opposite of 51
Opposites 194
Or 104, 108, 114
Order of a Radical 381
Order of Operations 26
Order of Radical 385
Ordered Pair 141, 147, 150
graph of 150
Ordered Triple 235, 238, 240
Ordinate 150
Origin 57, 150

P

Parabola 460, 461, 465, 470, 474
vertex 462, 466, 470, 475
Parallel Lines 185, 202
Percent 17, 137, 226
Perfect Powers 381
Perfect Square Trinomial 298
Perimeter 129, 218, 254
Perpendicular Lines 185
Placeholder Symbols 1
Polynomials 275, 278
adding 280
dividing 284
four-term 305
multiplying 282
subtracting 280
Power 263
Power Property of Logarithms 532
Prime Factored Form 20
Prime Number 20
Principal Root 377
Principle of Powers 395, 398
Product Property for Radicals 379
Product Property of Logarithms 532
Properties of Operations 29
Property
addition for radicals 387
additive inverse 33
associative 30
commutative 30
distributive 30, 87, 89
identity 32
logarithm of equality 546
multiplicative inverse 33
power of logarithms 532
product for radicals 379
product of logarithms 532
quotient for radicals 385
quotient of logarithms 532
zero-product 288, 349
Property of Equality
addition 60, 194
multiplication 60, 194
reflexive 60
symmetric 60
transitive 60
Property of Inequality
addition 63
multiplication 63
Pythagorean Theorem 403, 448

Q

Quadrant 150
Quadratic
discriminant 430
formula 426
reducing to 435, 438, 441
relations 419
roots of 419
standard form 419
Quotient Property for Radicals 385
Quotient Property of Logarithms 532

R

Radical
adding 387
addition property 387
compatible for addition and subtraction 387
compatible for division 385
compatible for multiplication 379, 381
division 385
equation 395, 398
expression 375
isolate 395, 398
multiplication 379
order of 381, 385
product property for 379
quotient property 385
sign 375
simplifying expressions 381
subtracting 387
Radicand 375
Radius of a Circle 482, 485
Rational
expression 331
Rational Exponent 375
Rational Numbers 8, 57
Rationalize a Denominator 392, 415
Real Numbers 8
Reciprocal 79, 83, 271
Rectangle of Reference 496, 502
Rectangular Coordinate System 149
Reflexive Property of Equality 60
Relation 70
Relation Symbols 1, 57, 60, 63
Relatively Prime Numbers 24
Repeating Decimal 12, 14
Repetend 12, 14
Restricted Values 331
Resultant Vector 44, 47, 53
Right Circular Cone 459
Root 375
cube 377
principal 377
square 377
Roots
cube 556
extraneous 395, 398
square 556
Roots of a Quadratic Relation 419
Roster 66
Rule 66

S

Scientific Notation 273, 535, 541
Set 66
finite 66
infinite 66
members of 66
null 66
solution 73
Significant Digits 273
Slope 153, 156, 164, 168, 172, 177, 179, 182
Slope-Intercept Form of an Equation 177, 179, 182
Solution Set 73
finding relations for 177, 179, 182, 185
for equations 141
for inequalities 147
graphing 153, 156
of absolute value 172
open sentence 149, 153, 156, 159
special cases 164, 168, 172
Solution Statement 73
Square Root 377
Standard Form of a Quadratic Relation 419
Subtraction
complex numbers 410
of fractions 338
radicals 387
Sum of Cubes 308
Supplementary Angles 129, 218, 254
Symbols 1
grouping 1, 87
number 1, 7, 10, 12, 14, 17, 20, 22, 24
operation 1, 26, 29, 32, 34, 37, 39, 41, 44, 47, 51, 53
placeholder 1
relation 1, 57, 60, 63
translating words into 5

Symmetric Property of Equality 60
Synthetic Division 315
System
  consistent 238, 244
  dependent 244
  inconsistent 242
  independent 238, 242
  of relations 506, 509
System of Simultaneous Linear
    Equations 188, 194, 199, 202, 206
System of Simultaneous Linear
    Inequalities 191

T

Table of Common Logarithms 557,
    558
Table of Square and Cube Roots 556
Term 276, 278
  degree of 276
Terminating Decimal 12, 14
Transitive Property of Equality 60
Trapezoid 129
Triangle 129
Trinomial 278
  general 301
  perfect square 298

U

Uniform Motion Relation 133, 222
Union 66

V

Variable
  degree of 276
Variation
  combined 371
  direct 366
  inverse 368
  joint 370
  mixed 370
Vector 44, 47, 53
Vector Addition 44, 47
Venn Diagram 67
Vertex 172
Vertex of a Parabola 462, 466, 470,
    475
Vertical Axis 150
Vertices of a Hyperbola 496, 502

W

Well Defined Operation 29
Well-Defined Operation 7
Whole Numbers 7, 57

Y

y-Intercept 177, 179, 182

Z

Zero-Product Property 288, 349